CELL BIOLOGY RESEARCH PROGRESS

PLATELETS

OVERVIEW, FUNCTION AND DISORDERS

CELL BIOLOGY RESEARCH PROGRESS

Additional books and e-books in this series can be found
on Nova's website under the Series tab.

CELL BIOLOGY RESEARCH PROGRESS

PLATELETS

OVERVIEW, FUNCTION AND DISORDERS

RENÉ LANGELIER
EDITOR

Copyright © 2019 by Nova Science Publishers, Inc.

All rights reserved. No part of this book may be reproduced, stored in a retrieval system or transmitted in any form or by any means: electronic, electrostatic, magnetic, tape, mechanical photocopying, recording or otherwise without the written permission of the Publisher.

We have partnered with Copyright Clearance Center to make it easy for you to obtain permissions to reuse content from this publication. Simply navigate to this publication's page on Nova's website and locate the "Get Permission" button below the title description. This button is linked directly to the title's permission page on copyright.com. Alternatively, you can visit copyright.com and search by title, ISBN, or ISSN.

For further questions about using the service on copyright.com, please contact:
Copyright Clearance Center
Phone: +1-(978) 750-8400   Fax: +1-(978) 750-4470   E-mail: info@copyright.com

NOTICE TO THE READER

The Publisher has taken reasonable care in the preparation of this book, but makes no expressed or implied warranty of any kind and assumes no responsibility for any errors or omissions. No liability is assumed for incidental or consequential damages in connection with or arising out of information contained in this book. The Publisher shall not be liable for any special, consequential, or exemplary damages resulting, in whole or in part, from the readers' use of, or reliance upon, this material. Any parts of this book based on government reports are so indicated and copyright is claimed for those parts to the extent applicable to compilations of such works.

Independent verification should be sought for any data, advice or recommendations contained in this book. In addition, no responsibility is assumed by the Publisher for any injury and/or damage to persons or property arising from any methods, products, instructions, ideas or otherwise contained in this publication.

This publication is designed to provide accurate and authoritative information with regard to the subject matter covered herein. It is sold with the clear understanding that the Publisher is not engaged in rendering legal or any other professional services. If legal or any other expert assistance is required, the services of a competent person should be sought. FROM A DECLARATION OF PARTICIPANTS JOINTLY ADOPTED BY A COMMITTEE OF THE AMERICAN BAR ASSOCIATION AND A COMMITTEE OF PUBLISHERS.

Additional color graphics may be available in the e-book version of this book.

Library of Congress Cataloging-in-Publication Data

ISBN: 978-1-53616-592-0

Published by Nova Science Publishers, Inc. † New York

CONTENTS

Preface		vii
Chapter 1	Platelet Functions *Elżbieta Lachert*	1
Chapter 2	The Platelets Role in Cutaneous Melanoma *Aline Mânica and Margarete Dulce Bagatini*	25
Chapter 3	Autologous Platelet-Rich Plasma (PRP): A Treatment for Difficult-To-Heal Venous Leg Ulcers *Valeria G. Mateeva, Doncho N. Etugov and Grisha S. Mateev*	43
Chapter 4	Thrombocytopenia: Emphasis on Etiology and Therapeutics *R. Vani and M. Manasa*	57
Index		111
Related Nova Publications		119

PREFACE

Hemostasis involves a set of strictly regulated processes that maintain blood in its fluid state within the vascular bed and, in case of vessel injury, promote the formation of platelet plugs and fibrin clots to prevent blood extravasation. As such, *Platelets: Overview, Function and Disorders* explores how platelets play an important role in the blood coagulation process, and how platelet deficiencies or functional defects may be the cause of some bleeding disorders.

The worldwide incidence of cutaneous melanoma has been increasing annually at a more rapid rate in comparison to any other type of cancer affecting predominantly young and middle-aged individuals. It is known that when melanoma cells leave the primary tumor and enter the blood stream, they activate surrounding platelets via some molecules, inducing microthrombus formation. The authors discuss how platelets contribute to inflammation, cancer invasion, and metastasis.

Additionally, the role of platelet-rich-plasma in stimulating the healing process in difficult-to-heal ulcers has been investigated over the past 25 years. It is suggested that platelet-rich-plasma is capable of transforming the difficult-to-heal skin ulcer with low metabolic activity into a healing ulcer with increased capacity for tissue regeneration.

This closing study focuses on the causes, diagnosis, and prognosis of various types of thrombocytopenia, providing an outline on the future prospects of using antioxidants for the treatment of a few thrombocytopenic conditions.

Chapter 1 - Hemostasis involves a set of strictly regulated processes that maintain blood in its fluid state within the vascular bed and in case of vessel injury promote formation of first platelet plug and then fibrin clot to prevent blood extravasation. Platelets play an important role in the blood coagulation process. Platelet deficiencies or functional defects may be the cause of some bleeding disorders. On the other hand, overproduction or activation of platelets may cause thromboembolic complications. The main role of platelets is their participation in primary hemostasis (adhesion, activation, release reaction and aggregation) and in blood clotting process through supply of phospholipids. Until recently, the traditional role of platelets has been thought to be limited to participation in hemostasis. However, deeper insight into platelet metabolism has demonstrated the potential role of platelets also in many other physiological and pathophysiological processes such as for example: transport, immunological processes, wound healing and inflammatory process. Platelets are involved in the transfer of certain substances during their flow in the circulatory system. On their surface they adsorb plasma coagulation factors, immunoglobulins, proteinase inhibitors and albumin. Platelets also participate in the processes of transplant rejection. As examples we may refer to renal allografts where platelet aggregation caused by endothelial injury or antigen-antibody complexes is found to be the cause of transplant rejection. Platelet gel which contains platelet-derived growth factors enhances the physiological response that occurs during wound healing. Animal studies have demonstrated that administration of platelet gel with high concentration of growth factors contributes to the healing and regenerating properties of the gel following complicated surgeries of both soft and hard tissues. Cooperation between platelets, leukocytes and endothelial cells and the ability to communicate cross-talk confirms the participation of platelets in inflammatory processes.

Chapter 2 - The worldwide incidence of cutaneous melanoma has been increasing annually at a more rapid rate compared to any other type of cancer affecting mostly young and middle-aged individuals (the average of 57 years at diagnosis). Over the past years, a deeper understanding of melanoma development and biology has been reached. It is known that when melanoma cells leave the primary tumor and enter the blood stream, they activate surrounding platelets via some molecules, inducing microthrombus formation. The platelets contribute to inflammation, cancer invasion, and metastasis. It was been demonstrated that tumor cells have the ability to induce platelet activation and aggregation and the cancer promotes platelet activation and activated platelets participate in each step of cancer progression. This knowledge has led to the identification of new therapeutic targets and treatment strategies.

Chapter 3 - Venous leg ulcers (VLUs) are a common medical problem in everyday practice. The condition is encountered in around 5% of the elderly above 65 years and in 1.5% of the general population. The management of VLUs is interdisciplinary and presents a significant economic burden. Furthermore, the importance of VLUs is increasing due to the worldwide demographic tendency for aging of the population. Conventional methods for treatment of VLUs include: compressive therapy, local wound care such as mechanical debridement and local antibiotics in case of infection; systemic medication and surgery. Often these methods require prolonged treatment and lack in efficacy. The fundamental pathophysiological process in VLUs is inflammation of the venous wall due to increased hydrostatic pressure in the vascular system of the lower extremities. The induced inflammatory response consists of interplay between inflammatory cells such as leukocytes, macrophages and monocytes, T-lymphocytes, and inflammatory molecules such as mediators, chemokines, growth factors, etc. Growth factors derived from the thrombocytes modify the microenvironment of the wound and stimulate the capillary angiogenesis, fibroblastic migration and proliferation, collagen synthesis and reepitelisation. The most important thrombocytic growth factors are: platelet-derived growth factor (PDGF), platelet-derived angiogenesis factor (PDAF), platelet-derived epidermal growth factor (PDEGF) and platelet factor 4. The role of the platelet-rich-plasma (PRP) in stimulating the healing process in difficult-to-heal ulcers has been investigated in the past 25 years. It is suggested that PRP is capable to transform the difficult-to-heal skin ulcer with low metabolic activity into a healing ulcer with increased capacity for tissue regeneration.

Chapter 4 - Platelets are anucleate cells, of 1-2 microns and are the second most abundant cells in blood. Their concentration in blood ranges between 150 - 450 x 10^9/L in an average adult human. Megakaryocytes are the precursors of platelets in bone marrow, from which proplatelets are released into the bloodstream due to shear stress. Humans produce ~10^{11} platelets per day, and with a life span of 7 – 10 days in circulation. Platelets have a major role in repairing the damaged endothelium of blood vessels. They arrest bleeding at the site of vascular injury through a process involving platelet adhesion, activation, secretion, and aggregation, subsequently leading to the formation of a hemostatic plug. Thrombocytopenia is a platelet disorder caused due to the reduction in platelet number (< 150 x 10^9 cells/L). It is further classified based on the platelet count, as mild (> 70 x 10^9 cells/L), moderate (20 - 70 x 10^9 cells/L) and severe (< 20 x 10^9 cells/L). Severe thrombocytopenia can be life-threatening. Thrombocytopenia can be attributed to factors such as (a) increased platelet destruction, (b) reduced platelet production and (c) abnormal platelet distribution/splenic pooling, or a combination of these factors. Thrombocytopenia can be additionally classified based on pathogenesis as:

1. Pseudo thrombocytopenia is due to *ex vivo* agglutination of platelets. This condition is observed when EDTA is used as an anticoagulant. However, it is necessary to confirm pseudo thrombocytopenia by manual examination of a blood smear to avoid unnecessary tests and treatment.
2. Thrombocytopenia due to reduced platelet production.
    (a) Inherited platelet disorders are caused due to genetic defects affecting the production of platelets, their functions, and morphology.
    (b) Acquired platelet disorders can be immune-mediated, drug-induced, pregnancy-related or due to nutritional deficiencies.
3. Thrombocytopenia due to elevated platelet destruction can be drug-induced, immune-mediated, or related to artificial surfaces

(like materials used during hemodialysis or cardiopulmonary bypass).
4. Thrombocytopenia due to abnormal platelet distribution can be caused by hypersplenism, hypothermia or transfusions.
5. Thrombocytopenia due to other causes include cyclic thrombocytopenia and acquired pure megakaryocytic thrombocytopenia.

Various causative factors form the basis of the strategies employed in thrombocytopenia treatments. Essentially, withdrawal of the drug causing thrombocytopenia is an important therapeutic measure in drug-induced thrombocytopenia. Treatment for immune thrombocytopenia includes glucocorticoids, intravenous immunoglobulins, splenectomy, anti-(Rh)D, thrombopoietin receptor agonists, drugs like rituximab, romiplostim, eltrombopag, azathioprine, cyclophosphamide, cyclosporine, vinca alkaloids, etc. However, treatment of the underlying disease, apart from treating thrombocytopenia, is recommended for patients with splenomegaly and viral infections. This chapter focuses on the causes, diagnosis, and prognosis of various types of thrombocytopenia. It also gives an outline on future prospects of using antioxidants for the treatment of a few thrombocytopenic conditions.

In: Platelets
Editor: René Langelier

ISBN: 978-1-53616-592-0
© 2019 Nova Science Publishers, Inc.

Chapter 1

PLATELET FUNCTIONS

*Elżbieta Lachert**
Department of Transfusion Medicine,
Institute of Hematology and Transfusion Medicine,
Warsaw, Poland

ABSTRACT

Hemostasis involves a set of strictly regulated processes that maintain blood in its fluid state within the vascular bed and in case of vessel injury promote formation of first platelet plug and then fibrin clot to prevent blood extravasation. Platelets play an important role in the blood coagulation process. Platelet deficiencies or functional defects may be the cause of some bleeding disorders. On the other hand, overproduction or activation of platelets may cause thromboembolic complications. The main role of platelets is their participation in primary hemostasis (adhesion, activation, release reaction and aggregation) and in blood clotting process through supply of phospholipids.

Until recently, the traditional role of platelets has been thought to be limited to participation in hemostasis.

---

* Corresponding Author's E-mail: elzbieta.lachert@gmail.com.

However, deeper insight into platelet metabolism has demonstrated the potential role of platelets also in many other physiological and pathophysiological processes such as for example: transport, immunological processes, wound healing and inflammatory process.

Platelets are involved in the transfer of certain substances during their flow in the circulatory system. On their surface they adsorb plasma coagulation factors, immunoglobulins, proteinase inhibitors and albumin. Platelets also participate in the processes of transplant rejection. As examples we may refer to renal allografts where platelet aggregation caused by endothelial injury or antigen-antibody complexes is found to be the cause of transplant rejection.

Platelet gel which contains platelet-derived growth factors enhances the physiological response that occurs during wound healing. Animal studies have demonstrated that administration of platelet gel with high concentration of growth factors contributes to the healing and regenerating properties of the gel following complicated surgeries of both soft and hard tissues. Cooperation between platelets, leukocytes and endothelial cells and the ability to communicate cross-talk confirms the participation of platelets in inflammatory processes.

Keywords: platelets, hemostasis

INTRODUCTION

Hemostasis involves a set of strictly regulated processes that maintain blood in its fluid state within the vascular bed and in case of vessel injury promote formation of first platelet plug and then fibrin clot to prevent blood extravasation. Until recently, the traditional role of platelets has been thought to be limited to participation in hemostasis, i.e., platelet plug formation and role in the blood clotting process. However, deeper insight into platelet metabolism has demonstrated the potential role of platelets also in many other physiological and pathophysiological processes such as for example: transport, immunological processes, in wound healing and in inflammatory process [1, 2].

ROLE OF PLATELETS IN PRIMARY HEMOSTASIS

Resting platelets are small discoid shapes around 2 to 4 µm in size with a diversified internal structure. Important for all platelet functions is the three-layer membrane and platelet granules (Table 1). Two layers of the membrane are phospholipids which contain cholesterol, glycolipids and glycoproteins. The platelet-membrane is responsible for transport, it controls the ionic environment and participates in the adhesion and release reaction. This is the location site for receptors of proteins involved in blood coagulation. These are single glycoproteins (GP) or their complexes. On the basis of the speed of migration in electrophoresis, they were given symbols: Ia, Ib, IIa, IIb, etc. About 50 glycoproteins have been detected in human platelets but the functions of only few of them have been identified. Glycoproteins involved in adhesion to components of connective tissue and in aggregation have been identified. Several glycoproteins may participate in direct adhesion of platelets to collagen (GP Ia/IIa, GP IIb/IIIa complexes). The GP Ia/IIa complex is also responsible for binding the platelets with laminin and vascular endothelium. GP Ic/IIa is the receptor for fibronectin. GP IIb/IIIa, a non-covalent complex consisting of two subunits IIb and IIIa plays a critical role in platelet aggregation. GP IIb is composed of an α chain with molecular weight of 120 kD and a β chain with a molecular weight of 25 kD, connected by the S-S bridge. A single GP IIIa chain has a molecular weight of 98 kD. In the resting platelet GP IIb/IIIa is recognized by antibodies on the surface of the cell membrane, as well as in the membranes of open canalicular system (OCS) and granules; it is the receptor for fibrinogen. GP IIb/IIIa of activated platelets also has the ability to bind von Willebrand factor (vWF), fibronectin, thrombosphodine and vitronectin which means that in cases such as afibrinogenemia, vWF may support platelet aggregation. The most important role of platelets is formation of a clot that prevents the outflow of blood from an injured vessel. There are several stages of this process, namely: platelet adhesion to subendothelial layer, platelet activation, platelet release reaction and platelet aggregation [2, 3].

Table 1. Content of platelet granules [44, 45, 46]

Types of granules	Content	
Alpha granule	Specific proteins (not present in plasma)	Chemokines: CXCL1 (beta – thromboglobulin, ß-TG), CXCL4 (platelet factor 4, PF4)
	Adhesive proteins	Fibronectin (obtained by endocytosis from plasma), vWF, Thrombospondin (synthesized in megakaryocytes)
	Active proteins in blood clotting and fibrinolysis	Fibrinogen, F.V, F.XI, high-molecular-weight kininogen, *C1*-esterase-*inhibitor, protein S,* activators and inhibitors of fibrinolysis
	Specific platelets mitogens	PDGF
	Proteins occurring exclusively in the alpha granule membrane	P selectin
	Proteins present in the alpha granule membrane and in the cytoplasmic membrane	GP IIb/IIIa, GP Ib/IX/V, GP IV
Dense granule	Adenosine diphosphate (ADP), adenosine triphosphate (ATP), guanosine diphosphate (GDP), guanosine triphosphate GTP, calcium ions, magnesium ions, 5-hydroxytryptamine (serotonin)	
	Phosphoinositols, polyphosphates,	
Lysosomes	Acid hydrolases	
Peroxisomes	Catalase	

ADHESION

Adhesion is the first stage of primary hemostasis. Following vascular endothelial injury platelets "bind" to adhesive proteins such as collagen, fibronectin, vitronectin and laminin of the subendothelial connective tissue. The subendothelial layer contains tissue factor (TF), i.e., CD 142, a protein found in the smooth-muscle membranes, fibroblasts and macrophages which initiate the blood coagulation process.

Vascular endothelial cells are the synthesis-site of numerous compounds relevant for hemostasis such as - among others - plasminogen activators and inhibitors and vW factor. In vessels where blood flow generates high shear stress (e.g., in small arteries) the binding of vWF to platelets and the subendothelial matrix is not very strong. The platelet receptor, responsible for binding with vWF is the surface glycoprotein complex GPIb/IX/V. In vessels where blood flow generates low shear stress, platelets bind directly to collagen via their GPIa/IIa complex. VWF is a component of the multimolecular factor VIII complex. VWF multimers bind with coagulation factor VIII (VIIIc; procoagulant in the intrinsic coagulation pathway). VWF combined with factor VIII:C extends the circulation time of factor VIII:C. These factors are controlled by different genes and are synthesized independently.

After binding with subendothelial connective tissue vWF changes shape which facilitates binding with the platelet GPIb receptor; especially under sheer stress when blood has contact with the subendothelial surface. GP Ib deficiencies or abnormalities are responsible for impaired platelet adhesion in the Bernard-Soulier's syndrome. This desease is responsible for severe lifelong bleeding disorders. GP Ib is also characterized by the ability to bind thrombin, which results in the displacement of GP Ib into the cell interior and decreases the exposure of this glycoprotein on the surface. Reduction of GP Ib expression also occurs as effect of other activating factors and serves as an indicator of the degree of platelet activation [3, 4, 5].

ACTIVATION

Adhesion of platelets to the proteins of subendothelial connective tissue as well as binding of collagen agonists with the receptors of the platelet membrane activates intracellular signals for activation. In vivo, platelet activation is initiated primarily by thrombin, collagen, adrenaline, platelet activating factor (PAF) and ADP. PAF is a phospholipid released from stimulated leukocytes and endothelial cells. Further activation and aggregation is stimulated by substances released from the platelets themselves: ADP, thromboxane A2 and serotonin [6]. Platelets then pass on from resting state to active state which manifests in a change of shape - from discoid to irregular with numerous pseudopodia. In the activated platelets there occurs translocation of proteins. Biologically active substances are released from the platelet granules and they enhance platelet activation. One of the effects is conformational changes of the GP IIb/IIIa which is the receptor for fibrinogen, the dimetric molecule of which binds two platelets. Inherited GP IIb/IIIa deficiencies, known as Glanzman's thrombasthenia, are responsible for lifelong bleeding episodes [7].

Platelet Release Reaction

An energy-dependent mechanism of contraction is required for platelet granules to be released or for degranulation to occur. As result, dense granules and alpha granules are crushed by the surrounding network of microtubules and microfilaments and their content is released into the open canalicular system that has connection with the platelet surface. Shrinkage of platelets and their concentration makes it easier to release the contents. Platelet aggregation is promoted by ADP, serotonin and calcium released from the granules [8, 9].

Aggregation

Adenosine 5-diphosphate (ADP), thromboxane A2, thrombin and PAF are important agonists that stimulate platelet aggregation. Fibrinogen and Ca ions also play a significant role in the aggregation process. Agonists are responsible for activation of the GPIb/IIIa complex, the platelet surface receptor for β3 integrin, which binds fibrinogen. Aggregation occurs when symmetrical fibrinogen molecules bind with the exposed receptors of adjacent platelets [10].

THE ROLE OF PLATELETS IN BLOOD COAGULATION

Platelets are not only active in hemostatic plug formation (primary hemostasis) but also participate in the process of blood clotting through supply of phospholipids. Such procoagulant activity is characteristic of negatively charged phospholipids such as phosphatidylserine and phosphatidylethanolamine, which in resting platelets are located in the inner layers of the cell membrane. Under the influence of activating agents, e.g., thrombin or collagen, phospholipids are displaced to the platelet surface. They form complexes with coagulation factors IXa/VIIIa/X (tenase) and with factors Va/Xa (prothrombinase), which are involved in the process of thrombin formation. An important role in the activation of coagulation is also attributed to the formation microparticles which are cleaved from activated platelets. Microparticles are spherical membrane fragments of 0.1-1 μm in size. The microparticles contain, among others platelet integrins, P-selectin and phosphatidylserine and have the ability to bind to the endothelium and activate coagulation, even on undamaged vascular wall. Factor V released from alpha granule is bound on the surface of the activated platelets and is the receptor for factor X. A significant role is also attributed to factor VIII released from the alpha granules the concentration of which may markedly increase as a result of platelet activation. Intraplatelet substances, such as P-selectin and CD40L released during platelet activation may intensify thrombin generation by stimulating TF synthesis in monocytes and endothelial cells [11, 12].

TRANSPORT FUNCTIONS OF PLATELETS

Platelets are involved in the transfer of certain substances during their flow in the circulatory system. On their surface they adsorb plasma coagulation factors, immunoglobulins, proteinase inhibitors and albumin. Biogenic amines are taken up by platelet cells on the principle of active transport with intact cell metabolism and stored in granules. In this way, serotonin and catecholamines are transported. In contrast, histamine penetrates into the platelet cell by way of diffusion.

One of the hypotheses is that serotonin and other biogenic amines that are transported by platelets may participate in the mechanisms of migraine headaches [11]. The etiopathogenesis of this disease still remains unclear, numerous data however point to the vascular nature of migraine seizures. It is likely that first the head vessels contract (aura) and then expand, which causes acute pulsating pain. It is presumed that these vascular symptoms may be caused by the release from platelets of such vasoactive substances, like e.g., serotonin, histamine and prostaglandin E1. Platelets of migraine sufferers have been shown to manifest aggregation disorders. Particularly intensive is the aggregation of such platelets under the influence of serotonin alone. It is also presumed that the platelets of migraine patients have poorer ability to capture serotonin from the environment [12].

THE ROLE OF PLATELETS IN SOME IMMUNOLOGICAL PROCESSES

Platelets also participate in the processes of transplant rejection. As examples we may refer to renal allografts where platelet aggregation caused by endothelial injury or antigen-antibody complexes is found to be the cause of transplant rejection. Platelet clusters and fibrin deposits in renal vessels lead to microthrombi formation. In the group of reversible processes of renal transplant rejection platelet plugs were observed in the capillaries of these organs; the platelets themselves were undamaged but with pseudopodia. They did not however lose their granule content. With intensive immunosuppressive therapy, renal function in most cases was resumed. In cases unsusceptible to treatment and requiring graft removal the renal capillaries contained degranulated platelets in which release reaction occurred and a hemostatic plug containing fibrin fibers was formed.

It is known for a fact that antigen-antibody complexes induce aggregation which trigger the clotting process. There exists a large group of immune-mediated kidney diseases. Fibrin deposits have been found in almost all forms of glomerulonephritis. Animal studies have demonstrated that when the reticuloendothelial system was blocked and infusion of soluble immune complexes took place there occurred massive deposition of fibrinogen derivatives which led to kidney necrosis.

THE ROLE OF PLATELETS IN WOUND HEALING

The goal of modern surgery is to achieve effective and rapid tissue reconstruction following surgical procedure, trauma or inflammation process. Although marked advancement has been achieved in contemporary surgical techniques there is still the ongoing search for ideal factors and agents to improve post-surgery hemostasis and wound healing. In this contex, platelet gels deserve closer attention [13, 14].

Already in the early 1980s, in vitro studies have shown that platelet-derived polypeptide growth factors regulate early proliferation and contribute to the differentiation and synthesis of all cell types that play an important role in the repair of both soft and hard tissues.

In 1987 and 1989, in clinical studies, the stimulating effects of growth factors in the healing process of tissues were confirmed. It was also found that the platelets contain human platelets of antimicrobial peptides (HPAPs), which in vitro microbiological tests have shown activity against some microorganisms, such as *Escherichia coli, Staphylococcus aureus, Cryptococcus neoformans.* This has stimulated growing interest in platelet products and their application in regenerative medicine [15].

Since the beginning of the 1990s numerous methods of obtaining platelet gels have been developed. Currently, under laboratory conditions it is possible to obtain autologous or allogeneic platelet-rich plasma which is the basic component of platelet gel.

Platelet gel is based on two components: platelet-rich plasma (PRP) or platelet concentrates (PCs) and thrombin solution. During mixing thrombin transforms fibrinogen into fibrin monomers and the thrombin-induced activation of coagulation Factor XIII in the presence of calcium ions induces the stabilization of cross links of fibrin monomers. Moreover, thrombin-induced platelet activation stimulates the release of numerous therapeutically active growth factors from platelets granules. Functions of growth factors in the process of healing and tissue reconstruction are presented in Table 2 [15, 16, 17].

Platelet gel which contains platelet-derived growth factors enhances the physiological response that occurs during wound healing. Animal studies have demonstrated that administration of platelet gel with high concentration of growth factors contributes to the healing and regenerating properties of the gel following complicated surgeries of both soft and hard tissues. Platelet derived growth factor (PDGF) is particularly important for hard-healing wounds, acts as stimulator of revascularization, collagen synthesis and bone regeneration.

Table 2. Major platelet's growth factors and their roles [15]

Growth Factor	Main role
Epidermal Growth Factor (EGF)	stimulates angiogenesis stimulates proliferation of myoblasts mitogenic for mesenchymal cells, chondrocytes and osteoblasts promotes growth/differentiation of chondrocytes and osteoblasts
Fibroblast Growth Factor (FGF)	enhances endothelial cell proliferation and promotes physical organization of endothelial cells into tube-like structures promotes angiogenesis stimulates proliferation of myoblasts.
Platelet Derived Growth Factor (PDGF a-b)	stimulates cell replication and angiogenesis regulates collagen synthesis mitogenic for fibroblast/glial/smooth muscle cells, mesenchymal cells and osteoblasts. stimulates macrophage and neutrophil chemotaxis
Transforming Growth Factor (TGF-α, TGF-β)	basic growth and differentiation factor, involved in the healing of connective tissue and bone regeneration inhibits macrophage proliferation regulates mitogenic effects of other growth factors
Insulin-like Growth Factor (IGF)	stimulates keratenocyte proliferation, participates in the synthesis and proliferation of fibroblasts. acts on osteoblasts and their precursors
Vascular Endothelial Growth Factor (VEGF)	stimulates angiogenesis increases vessel permeability mitogenic for endothelial cells

Its role in the process of wound healing consists in stimulation of mitogenesis which increases the number of repair cells, stimulation of angiogenesis, which supports the development of new vessels and activation of macrophages responsible for wound cleansing and as secondary source of growth factors. Initially, platelet gels were used as support in cardiac surgery. Platelet gel was demonstrated to accelerate bone regeneration, reduce inflammation and blood loss and improve wound healing. Platelet gel has been used in complete reconstruction of hip and shoulder joints, fractures and bone defects, vertebral growth, clavicle resection or artroscopy. In addition, platelet gel enhances stabilization of soft tissue grafts and accelerates bone regeneration. It reduces postoperative bleeding, bruising, inflammation, swelling and post-operative drainage. For this reason, its use is recommended in plastic surgery with special emphasis on facial, dental and maxillo-facial surgery. Recently, it has been inreasingly applied in other surgical disciplines such as ENT surgery and trauma-orthopedic surgery. Routinely it has been used for many years now as supportive treatment for leg ulcers and diabetic foot ulcers [18, 19, 20].

ROLE OF PLATELETS IN INFLAMATORY PROCESSES

Currently, there is a growing interest in the role of platelets in inflammatory processes which engage the innate/immune system. Platelets interact with leukocytes and vascular endothelial cells and these processes may occur by way of immunological mediators secreted during the ongoing process. Long-term interaction of platelets with cells involved in the inflammation process may for example, reduce infection [21, 22].

During inflammation platelets secrete cytokines the role of which is to regulate the processes of immune response (e.g., proliferation and differentiation of lymphocytes) as well as participation in the process of hematopoiesis. Well known and well described inflammatory mediators include pro-inflammatory cytokines - interleukin 1 (IL-1) and tumor necrosis factor alpha (TNF-alpha). Cytokines that have chemotactic activity are called chemokines. Both cytokines and chemokines involved in the inflammatory process are stored in platelet granules (Table 1) [23, 24]. Toll Like Receptors (TLRs), present on neutrophils and macrophages, also play an important role in the inflammatory process. TLR1-TLR9 receptors are present on platelets [25, 26]. Recognition of the infectious agent by these receptors leads to the activation of the innate immune response. The relationship between TLR4 expression and the severity of thrombocytopenia induced by bacterial infection has been confirmed [27, 28]. It is also known from experimental studies that platelets bind circulating bacteria, which are then presented to neutrophils. In the course of the inflammatory process, the formation of cell microparticles is particularly intensive. Activated platelets are the source of platelet microparticles (PMPs) on the lipid membrane of which there are characteristic glycoprotein platelet markers [29, 30, 31]. PMPs more often bind to granulocytes and lymphocytes in which they induce adhesive molecule expression, phagocytic activity, cytokine secretion and affect angiogenesis. PMPs have also been found to participate in pathogenesis of certain autoimmune diseases. such as thrombocytopenia, lupus erythematosus or rheumatoid arthritis. The higher number of PMP may correlate with atherosclerosis in the course of diabetes mellitus, myocardial infarction, coronary artery disease and strokes [32, 33, 34, 35].

Mechanism of Platelet Participation in Inflammatory Processes

During the inflammatory process, there occurs interaction between leukocytes, endothelial cells and platelets. Activated endothelial cells show greater expression of adhesion molecules that facilitate rolling leukocytes to pass through the endothelial barrier of the vessel and reach the inflammation site. It has been demonstrated that the leukocyte-platelet complexes help neutrophils to pass through such a barrier [36].

Under physiological conditions, platelet cells do not react with the vascular endothelium. Only damage to the endothelium induces immediate adherence of platelets and their aggregation at the inflammation site. As in the case of leukocytes, there occurs the rolling of platelets into endothelial cells. E-selectin and P-selectin are essential for this process. In addition, platelets release soluble inflammatory mediators that affect the endothelium, while factors of inflammation such as IL-1 and TNF alpha induce the secretion of platelet chemokines involved in the passage of leukocytes through the endothelium [37, 38, 39].

In circulation activated platelet cells form leukocyte-platelet aggregates. Adhesion of these cells depends on platelet P-selectin and leukocytic P-selectin glycoprotein ligand 1(PSGL-1). The density of leukocyte platelet aggregates increases at the vessel circumference which enhances the risk of pathological deposits and atherosclerosis. Platelets and microparticles may affect adhesion of leukocytes in the inflammatory endothelium, which enhances leukocytes adhesion. This cooperation between platelets, leukocytes and endothelial cells and the ability to communicate cross-talk confirms the participation of platelets inflammatory processes [40, 41, 42, 43].

REFERENCES

[1] Smyth, S.S., Whiteheart, S., Italiano, J.E., Coller, B.S. (2010). *Platelet morphology, biochemistry and function. Williams Hematology.* Kaushansky, K., Beutler, E., Seligsohn, U., Lichtman, M.A., Kipps, T.J., Prchal, J.T., (red.). McGraw-Hill, New York: 1735–1814.

[2] Kopec, M. (1996). *Physiological Hemostasis*; Chapter 1; Thrombi and Emboli; (red.). Lopaciuk, S., Warsaw, PZWL.

[3] Boettiger, D. (2012). Mechanical control of integrin-mediated adhesion and signaling. *Curr. Opin. Cell Biol.,* 24: 592–599.

[4] Andrews, R. K., Gardiner, E. E., Berndt, M. C. (2003). Glycoprotein Ib-IX-V. *Int. J. Biochem. Cell Biol.,* 35:1170–1174.

[5] Chen, J., López, J.A. (2005). Interactions of platelets with subendothelium and endothelium. *Microcirculation,* 12: 235–246.

[6] Viisoreanu, D., Adrian, G. (2007). Effect of physiologic shear stresses and calcium on agonist-induced platelet aggregation, secretion, and thromboxane A_2 formation. *Thrombosis Research,* 120(6): 885-892.

[7] Nurden, A.T., Fiore, M., Pillois, X. (2011). Glanzmann thrombasthenia: a review of ITGA2B and ITGB3 defects with emphasis on variants, phenotypic variability, and mouse models. *Blood,* 118: 5996–6005.

[8] Rendu, F., Brohard-Bohn, B. (2001). The platelet release reaction: granules' constituents, secretion and functions. *Platelets,* 2001, 12: 261–273.

[9] Ren, Q., Ye, S., Whiteheart, S. W. (2008). The platelet release reaction: just when you thought platelet secretion was simple. *Curr Opin Hematol,* 15 (5): 537-41.

[10] Jackson, S.P. (2008). The growing complexity of platelet aggregation. *Blood*, 109 (12): 5087-5095.
[11] Gupta, S., Nahas, S.J, and Peterlin, B.L. (2011). Chemical Mediators of Migraine: Preclinical and Clinical Observations, *Headache*, 51(6): 1029-1045.
[12] Kotelba-Witkowska, B. (1984), *Plateles*. PZWL, Warsaw.
[13] Giusti, I., Rughetti, A., D'Ascenzo, S., Millimaggi, D., et al. (2009). Identification of an optimal concentration of platelet gel for promoting angiogenesis in human endothelial cells. *Transfusion*, 49: 771-778.
[14] Kazakos, K., Lyras, D.N., Verettas, D. et al. (2009). The use of autologous PRP gel as an aid in the management of acute trauma wounds. *Injury*, 8:801-805.
[15] Lachert, E. (2009), Fibrin Glue and Platelet Gel, *Transfusion Medicine*, (red.) J. Korsak and M. Letowska, α-medica press 2009.
[16] Burnouf, T. (2013). *Platelet gel.* International Society of Blood Transfusion, 8: 131-136.
[17] Findikcioglu, F., Findikcioglu, K., Yavuzer, R., Lortlar, N., Atabay, K. (2012). Effect of intraoperative platelet-rich plasma and fibrin glue application on skin flap survival. *The Journal of Craniofacial Surgery*, 23 (5): 1513-7.
[18] Wu, X., Ren, J., Yao, G., Zhou, B. et al. (2014). Biocompatibility, biodegradation, and neovascularization of human single-unit platelet-rich fibrin glue: an in vivo analysis. *Chinese Medical Journal*, 127 (3): 408-11.
[19] Asadi, M., Alamdari, H. D., Rahimi R. H, et al. (2014). Treatment of life-threatening wounds with a combination of allogenic platelet-rich plasma, fibrin glue and collagen matrix, and a literature review. *Experimental and therapeutic Medicine*, 8: 423-429.
[20] Ficarelli, E., Bernuzzi, G., Tognetti, E., Bussolati, O., et al. (2009). Treatment of chronic venous leg ulcers by platelet gel. *Dermatologic Therapy*, 21:S13-S17.
[21] Projahn, D., Koenen, R. R., (2012). Platelets: key players in vascular inflammation. *J. Leuk. Biol.*: 92, 1–9.

[22] Semple, J. W., (2012), Platelet have a role as immune cells. *ISBT Science Series,* 7: 269–273.
[23] Li, C., Li, J., Li Y. and et.(2012). Crosstalk between Platelets and the Immune System: Old Systems with New Discoveries. *Advances in Hematology,* Article ID 384685.
[24] Warren, J. S., Ward, P. A., (2010). *The inflammatory response. W: Williams Hematology,* Kaushansky K., Beutler E., Seligsohn U., Lichtman M. A., Kipps T. J., Prchal J. T. (red.). McGraw-Hill, New York.
[25] Maslanka, K., (2014). The role of platelets in inflammatory processes *Journal of Transfusion Medicine,* 7, 3: 102-109.
[26] Shiraki, R., Inoue, N., Kawasaki, S., et al. (2004). Expression of Toll-like receptors on human platelets. *Thromb. Res.,* 113: 379–385.
[27] Cognasse, F., Hamzeh, M., Chavarin, P., et al, (2005). Evidence of Toll-like receptor molecules on human platelets. *Immunol. Cell. Biol.,* 88: 196–198.
[28] Semple, J. W., Aslam, R., Kim, M., Speck, E. R., Freedman, J., (2007). Platelet-bound lipopolysaccharide enhances Fc receptor-mediated phagocytosis of IgG opsonize platelets. *Blood,* 109: 4803–4805.
[29] Aslam, R., Speck, E. R., Kim, M., et al. (2006). Platelet Toll-like receptor expression modulates lipopolysaccharide-induced thrombocytopenia and tumor necrosis factor-a production in vivo. *Blood,* 107: 637–641.
[30] Simak, J., Gelderman, M.P., (2006). Cell membrane microparticles in blood and blood products: potentially pathogenic agents and diagnostic markers. *Trans. Med. Rev.,* 20: 1–26.
[31] Piccin, A., Murphy, W. G., Smith, O. P., (2007). Circulating microparticles: pathophysiology and clinical implications. *Blood,* 21: 157–171.
[32] Maslanka, K. (2010), Physiopathological activity of cell membrane microparticles. *J. Transf. Med.,* 1: 9–17.
[33] Semple, J. W., Provan, D., Garvey, M. B., Freedman, J., (2010), Recent progress in understanding the pathogenesis of immune thrombocytopenia (ITP). *Curr. Opin. Haematol.,* 17: 590–595.

[34] Nagahama, M., Nomura, S., Ozaki, Y., et al. (2001). Platelet activation markers and soluble adhesion molecules in patients with systemic lupus erythematosus. *Autoimmunity*, 33: 85–94.
[35] Boilard, E., Nigrovic, P.A., Larabee, K., et al. (2010). Platelets amplify inflammation in arthritis via collagen-dependent microparticle production. *Science*, 327: 580–583.
[36] Diacovo, T. G., Roth, S. J., Buccola, J. M., et al. (2006). Neutrophil rolling, arrest, and transmigration across activated, surface-adherent platelets via sequential action of P-selectin and the beta 2-integrin CD11/CD18. *Blood*, 88: 146–157.
[37] Kopec-Szlezak, J., (2014), The migration of hematopoietic cells and leukocytes, *J. Transf. Med.*, 7: 40–50.
[38] Frenette, P. S., Denis, C. V., Weiss, L., et al. (2000). P-Selectin glycoprotein ligand-1 (PSGL-1) is expressed on platelets and can mediate platelet-endothelial interactions in vivo. *J. Exp. Med.*, 191: 1413–1422.
[39] Romo, G. M., Dong, J. F., Schade, A. J., et al, (1999). The glycoprotein Ib-IX--V complex is a platelet counter receptor for P-selectin. *J. Exp. Med.*, 190: 803–814.
[40] Huo, Y., Schober, A., Forlow, S. B., et al. Circulating activated platelets exacerbate atherosclerosis in mice deficient in apolipoprotein. *E. Nat. Med.*, 9: 61–67.
[41] Larsen, E., Celi, A., Gilbert, G. E., et al. (1989). PADGEM protein: A receptor that mediates the interaction of activated platelets with neutrophils and monocytes. *Cell*, 59: 305–312.
[42] Battrum, S. M., Hatton, R., Nash, G. B., (1993). Selectin-mediated rolling of neutrophils on immobilized platelets. *Blood*, 82: 1165–1174.
[43] Stokes, K. Y., Granger, D. N., (2012). Platelets: a critical link between inflammation and microvascular dysfunction. *J. Physiol.*, 590: 1023–1034.
[44] Golab, J., Jakobisiak, M., Firczuk, M., Lasek, W., Stoklosa, T. (red.) (2013). *Immunology*, Polish Scientific Publishers PWN, Warsaw: 157–197.

[45] Rossi, D., Zlotnik, A., (2000). The biology of chemokines and their receptors. *Ann. Rev. Immunol.,* 18: 217–202.
[46] Zlotnik, A., Yoshie, O., (2000). Chemokines: A new classification system and their role in immunity. *Immunity,* 12: 121–128.

BIOGRAPHICAL SKETCH

Elżbieta Lachert

Affiliation: Institute of Hematology and Transfusion Medicine (IHTM), Department of Transfusion Medicine

Education:

1981- Master of Pharmacy, Faculty of Pharmacy, Medical Academy in Warsaw,
1986 - specialist in laboratory diagnostics
1996 - Officer of Radiation Protection (type B)
2000 - Doctor of pharmacy, Faculty of Pharmacy, Medical Academy, PhD dissertation: *"Influence of gamma irradiation on platelets"*
2007 - specialist in laboratory transfusion medicine
2017 - title of doctor *habilitatus* and post of professor at the Institute of Hematology and Transfusion Medicine in Warsaw.

Research and Professional Experience:

Main fields of interest:

- Optimalization of methodology for autologous fibrin glue and platelet gel preparation
- Irradiation of blood components for prevention of TA- GvHD
- Platelet metabolism during storage of platelet concentrates

- Pathogen inactivation of labile blood products
- Safety of blood and blood components
- Methods of isolation and storage of stem cells

Scientific achievements:

- Author of 59 original papers and book chapters in Polish publicationsAuthor of 100 presentations at national and international conventions.
- Co-author of recommendations for the Polish blood establishments based on the regulations and recommendations of the European Union Directives.
- Representative of Competent Authorities as auditor for 23 Polish blood establishments and cell and tissue banks.
- Lecturer on topics related to transfusion medicine for post-graduate students
- Lecturer at conferences organized by scientific societies and the Institute of Hematology and Transfusion Medicine.

Professional Appointments:

- May 2001 – January 2008: holds the research and teaching post of *adiunk*t in the Laboratory of Quality Control, head of Laboratory of Quality Control, Department of Transfusion Medicine and Organization of Blood Transfusion Service,
- Since 2006 – chief consultant for the National Center for Tissue and Cell Banking
- January 2008- May 2011: head the Department of Quality Assurance and Organization of Blood Transfusion Service at IHTM
- March 2013 – August 2016: interim deputy director in Department of Transfusion Medicine at IHTM

- Since June 2011: head of the Laboratory of Quality Assurance at Department of Transfusion Medicine at IHTM,
- Since 2017 – professor of IHTM

Honors:

- In 2006 - award of the Institute of Hematology and Transfusion Medicine for : Medical standards for collection, preparation and distribution of blood and blood components in public blood transfusion establishments (Medyczne zasady pobierania krwi, oddzielania jej składników i wydawania obowiązujące w jednostkach organizacyjnych publicznej służby krwi). Antoniewicz-Papis J., Brojer E., Dzieciątkowska A., Kubis J., Kuśnierz-Alejska G., Lachert E., Maślanka K., Michalewska B., Mikulska M., Rosiek A., Seyfried H., Żupańska B.
- In 2006 - honorary award of the Ministry of Health for merits in the field of health protection – overall activity related to public blood transfusion service.
- In 2015 - collective award of the Main Board of Polish Transplantation Society for implementation of European guidelines for banking cells and tissue in Poland.

Publications from the Last 3 Years:

2016

1. Letowska, M., Przybylska, Z., Piotrowski, D., Lachert, E., Rosiek A., Rzymkiewicz, L., Cardoso, M. (2016). Hemovigilance survey of pethogen-reduced blood components in the Warsaw Region in the 2009 to 2013 period. *Transfusion,* 56, S: 39-44.
2. Lachert, E., Plodzich, A., Antoniewicz-Papis, J., Letowska, M. (2016). Sprawozdanie z udziału Instytutu Hematologii i Transfuzjologii w realizacji programu działań Unii Europejskiej w

dziedzinie zdrowia (2014-2020) Joint Action 8- informacje pilotażowe. *Journal of Transfusion Medicine,* 9: 61-71.
3. Rosiek, A., Tomaszewska, A., Lachert, E., Antoniewicz-Papis, J., Kubis, J., Pogłód, R., Letowska, M. (2016). Działalność jednostek organizacyjnych służby krwi w Polsce w 2015 roku. *Journal of Transfusion Medicin*e, 9 (4): 107-124.
4. Rosiek, A., Pogłód, R., Lachert, E., Antoniewicz-Papis, J., Rzymkiewicz, Letowska M. (2016). Therapeutic apheresis at the Institute of Hematology and Transfusion Medicine in Warsaw – a glimpse of recent tendencies, Vox Sanguinis 11, Suppl. 1, 250, P-455. *34th International Congress of the International Society of Blood Transfusion Dubai.*
5. Antoniewicz-Papis, J., Lachert E., Rosiek, A., Ejduk, A., Pienko, K., Letowska M. (2016). Retrospective validation of processes in routine use for autologous peripheral blood stem cell preparation and storage, Vox Sanguinis, 111, Suppl. 1, 283, P-542. *34th International Congress of the International Society of Blood Transfusion.* Dubai,
6. Lachert, E., Antoniewicz-Papis, J., Plodzich A., Gawrys, M., Kubis J., Rosiek, A., Przybylska, Z., Piotrowski, D., Łętowska M. (2016). Validation of Reveos TM system used for routine processing of blood components, Vox Sanguinis 2016, 111, Suppl. 1, 148, P-172. *34th International Congress of the International Society of Blood Transfusion Dubai.*

2017

1. Lachert, E., Wozniak, J., Antoniewicz-Papis, J., Krzywdzinska, A., Kubis, J., Mikolowska, A., Letowska M. (2017). Study of CD69 antigen expression and integrity of leukocyte cellular membrane in stored platelet concentrates following irradiation and treatment with Mirasol® PRT System *Adv Clin Exp Med,* 26: 7–13.
2. Cohn, C. S., Dumont, L. J., Lozano, M., Marks, D. C., Johnson, L. S., Bondar, I.N., Sas, F.T., Yokoyama, A. P. H., Kutner, J. M.,

Acker, J.P., Bohonek, M., Sailliol, A., Martinaud, C., Poglod, R., Antoniewicz-Papis, J., Lachert, E., Pun, P. B L, Lu, J., Cid, J, Guijarro, F., Puig, L., Gerber, B., Alberio,L., Schanz, U., Buser, A., Noorman, F., Zoodsma, M., van der Meer, P.F., de Korte, D., Wagner, S., & O'Neill, M. (2017). Vox Sanguinis International Forum on platelet cryopreservation: Summary. *Vox Sang*, 112: 684–688.
3. Antoniewicz-Papis, J, Lachert, E., Janik, K., Letowska, M. (2017). Autologous artificial tears used for treatment of dry eye syndrome in patients with chronic graft versus host disease *Pol Arch Intern Med,* 127 (10): 705-707.
4. Rosiek, A, Tomaszewska, A., Lachert, E., Antoniewicz-Papis, J., Kubis, J., Poglod, R., Letowska, M. (2017). Działalność jednostek organizacyjnych służby krwi w Polsce w 2016 roku. *J Transf Med.,* 10 (4): 113 – 129. [Activities of blood service organizational units in Poland in 2016]
5. Lachert, E., Antoniewicz-Papis, J. (2017). Wybrane zagadnienia dotyczące inaktywacji biologicznych czynników chorobotwórczych w składnikach krwi w świetle doniesień prezentowanych na *27. regionalnym Kongresie ISBT w Kopenhadze* (17–21.06.2017 r.), *J Transf Med.,* 10: 107-111. [Selected issues regarding the inactivation of biological pathogens in blood components in the light of reports presented at the *27th ISBT Regional Congress in Copenhagen*]
6. Mikolowska, A., Antoniewicz-Papis, J., Lachert, E. (2017),., *Wdrażanie nowych metod w krwiodawstwie, Laboratorium – Przegląd ogólnopolski*, 7-8: 18-22. [*Implementation of new methods in blood donation, Laboratory - Poland-wide review*]
7. Lachert, E. (2017). Zmiany w badaniach kontroli jakości składników krwi w Polsce. *Acta Haematol Pol:* 202-204. [Changes in blood component quality control tests in Poland.]
8. Lachert, E. (2017). Niekonwencjonalne zastosowanie składników krwi w lecznictwie. *Diagnostyka Laboratoryjna,* 53: 32-33.

[Unconventional use of blood components in medicine. Laboratory diagnostics]
9. Lachert, E. (2017). *Pathogen inactivation methods for blood and blood components*, Warsaw 2017, ISBN 978-83-947683-0-0.
10. Rosiek, A., Tomaszewska, A., Lachert, E., Antoniewicz-Papis, J., Poglod, R., Letowska, M. (2017). *Zasadnicze czynniki wpływające na zmienność stanu zasobów składników krwi na terenie Polski*. Acta Haemat. Pol., 48, Suppl. 1: 136. XXVII Zjazd Polskiego Towarzystwa Hematologów i Transfuzjologów. [*Essential factors affecting the variability of the state of blood components in Poland.*]
11. Poglod, R., Rosiek, A., Lachert, E., Grabarczyk, P., Michalewska, B., Łętowska, M. (2017). Analiza poważnych niepożądanych zdarzeń i poważnych niepożądanych reakcji poprzetoczeniowych w Polsce w latach 2011-2014. *XXVII Zjazd Polskiego Towarzystwa Hematologów i Transfuzjologów*, 21-23 września 2017 r. Warszawa, *Acta Haematol Pol,* 48, Supl 1:91. [Analysis of serious adverse events and serious adverse transfusion reactions in Poland in 2011-2014. *27th Congress of the Polish Society of Hematologists and Transfusion Medicine*]

2018

1. Debska, M., Uhrynowska, M., Guz, K., Kopec, I., Lachert, E., Orzinska, A., Kretowicz, P., Antoniewicz-Papis, J., Debski R., Letowska, M., Husebekk, A., Brojer, E. (2018). Identification and follow-up of pregnant women with platelet-type human platelet antigen (HPA)-1bb alloimmunized with fetal HPA-1a. *Arch Med Sci* 14, 5: 1041–1047.
2. Lachert, E., Kubis, J., Antoniewicz-Papis, J., Rosiek, A., Wozniak, J., Piotrowski, D., Przybylska. Z., Mikolowska, A., Marschner, S., Letowska, M. (2018). Quality control of riboflavin-treated platelet concentrates using Mirasol®PRT system: Polish experience. *Adv Clin Exp Med.*, 27(6): 765–772.

3. Lachert, E., Kubis, J., Antoniewicz-Papis, J., Bubinski M., M., Letowska M. (2018). Metody inaktywacji biologicznych czynników chorobotwórczych w koncentracie krwinek czerwonych i krwi pełnej. *J. of Transf. Med.* 11: 63-71. [Methods for inactivation of biological pathogens in concentrate of red blood cells and whole blood.]
4. Poglod, R., Rosiek, A., Michalewska, B., Lachert, E., Grabarczyk, P, Uhrynowska, M., Letowska, M.(2018). Analiza poważnych niepożądanych zdarzeń i poważnych niepożądanych reakcji poprzetoczeniowych w Polsce w latach 2011–2014. Część I. Poważne zdarzenia niepożądane i reakcje poprzetoczeniowe związane z obcogrupowym przetoczeniem składnika krwi. *Journal of Transfusion Medicin,* 11(1):8-28. [Analysis of serious adverse events and serious adverse transfusion reactions in Poland in 2011–2014. Part I. Serious adverse events and transfusion reactions associated with transgroup transfusion of the blood component]
5. Poglod, R., Rosiek, A., Michalewska, B., Kubis, J., Grabarczyk, P., Uhrynowska, M., Lachert, E., Letowska, M. (2018). Analiza poważnych niepożądanych zdarzeń i poważnych niepożądanych reakcji poprzetoczeniowych w Polsce w latach 2011–2014. Część II. Reakcje poprzetoczeniowe niezwiązane z obcogrupowym przetoczeniem składników krwi. *Journal of Transfusion Medicine,* 11 (3): 75-90. [Analysis of serious adverse events and serious adverse transfusion reactions in Poland in 2011–2014. Part II Transfusion reactions not related to transgroup transfusion of blood components]
6. Rosiek, A., Tomaszewska, A., Lachert, E., Antoniewicz-Papis, J., Kubis,J., Poglod, R., Letowska, M. (2018). Działalność jednostek organizacyjnych służby krwi w Polsce w 2017 roku. *Journal of Transfusion Medicine,* 11 (4): 113 – 130. [Activities of blood organizational units in Poland in 2017]

7. Kubis, J., Lachert, E., Antoniewicz-Papis, J. Rosiek, A., Mikolowska, A., Letowska M. (2018). Kontrola Jakości składników krwi w Centrach Krwiodawstwa i Krwiolecznictwa kontrolowanych w latach 2015-2016. *Journal of Transfusion Medicin,* 11 (4): 131 – 139. [Quality control of blood components in Blood Donation and Blood Treatment Centers controlled in 2015-2016.]

In: Platelets
Editor: René Langelier

ISBN: 978-1-53616-592-0
© 2019 Nova Science Publishers, Inc.

Chapter 2

THE PLATELETS ROLE IN CUTANEOUS MELANOMA

Aline Mânica[1], MD and
Margarete Dulce Bagatini[2,*], PhD

[1] Department of Biochemistry and Molecular Biology, Federal University of Santa Maria, Santa Maria, RS, Brazil.
[2] Academic Coordination, Campus Chapecó, Federal University of Fronteira Sul, Chapecó, Brazil

ABSTRACT

The worldwide incidence of cutaneous melanoma has been increasing annually at a more rapid rate compared to any other type of cancer affecting mostly young and middle-aged individuals (the average of 57 years at diagnosis). Over the past years, a deeper understanding of melanoma development and biology has been reached. It is known that when melanoma cells leave the primary tumor and enter the blood stream, they activate surrounding platelets via some molecules, inducing microthrombus formation. The platelets contribute to inflammation, cancer

---

* Corresponding Author's E-mail: margaretebagatini@yahoo.com.br.

invasion, and metastasis. It was been demonstrated that tumor cells have the ability to induce platelet activation and aggregation and the cancer promotes platelet activation and activated platelets participate in each step of cancer progression. This knowledge has led to the identification of new therapeutic targets and treatment strategies.

INTRODUCTION

Melanoma is a highly-malignant tumor which occurs in skin, mucosa, and visceral organs being the deadliest of all the skin cancers. It accounts for 4% of all skin cancers but nearly 75% of skin tumors-related mortality. The melanoma incidence has been increasing in recent decades, therefore providing early diagnosis and determining available and reliable biomarkers that can predict patient's prognosis are also an effective measure of improving survival (Leonardi et al., 2018).

The platelets are increasingly being recognized as a critical agent of immune function, and play a pivotal role in cancer growth, proliferation, invasion and metastasis. High platelet count has been promoting the progression of ovarian cancer, non-small cell lung cancer, pancreatic cancer, bladder cancer, gastrointestinal cancer and other tumors, which is also closely related to tumor prognosis (Qi et al., 2018).

The purpose of this chapter was to show the role of platelets in cancer metastasis and suppression of antitumor immunity, also supporting that targeting platelets can be used as adjuvant to immunotherapy in melanoma patients.

CUTANEOUS MELANOMA

Melanoma is often considered one of the most aggressive and treatment-resistant human cancers. Melanomas can arise within any anatomic territory occupied by melanocytes (Tsao et al., 2012) . Melanocytes are cells originated from the neural crest. Their migration to the epidermis, function, and their survival are all dependent on the expression of the tyrosine kinase receptor c-kit gene. Their main function is to produce melanin pigment in their specific cytoplasmic organelles called melanosomes (Bandarchi et al., 2010).

Melanin pigment synthesized by each melanocyte is transferred to an average of 36 keratinocyte then forms a cap at the top of the nucleus of mitotically active basal cells and prevents the ultraviolet injurious effects on the nucleus (Bandarchi et al., 2010). The melanin pigments can be divided into two main types: eumelanin and pheomelanin. The eumelanins are the black-brown subgroup of insoluble melanin pigments. The pheomelanins are yellow-to-reddish brown subgroup of melanin pigments (d'Ischia et al., 2015).

Melanocytes clustered together form nevi, which can vary in color in different skin tones. The nevi can be classified into acquired and congenital and can also be located in the epidermis and/or in the dermis. There are many studies supporting clinical and/or histopathological association of primary melanoma with melanocytic nevus (Saida 2019; WallaCeH. Clark, Jr., Lynn From, Evelina A. Bernardino 1969).

Acquired nevus appears during childhood and congenital nevus is defined as melanocytic nevus seen at or soon after the birth of infants. Both have very small size to be considered a benign neoplasm of epidermal melanocyte, but they could contribute to melanoma development (Saida 2019). Dysplastic nevus is considered an intermediate lesion between benign nevus and melanoma lesion. According to the latest version of the WHO Classification of Skin Tumors published in 2018, dysplastic nevi are a subset of melanocytic nevi that are clinically atypical and characterized histologically by architectural disorder and cytological atypia, always involving their junctional component.

Cutaneous melanomas are visible to the eye and represent an apparent disease. Although exposure to ultraviolet (UV) radiation and rare genetic susceptibility within some ethnic groups are associated with the development and progression of melanoma, very little is known about the etiology of this tumor (Gil et al., 2019). About the risk factors to melanoma development, nowadays it is considered as a multi-factorial disease.

Some important risk factors include: exposure to UV rays because of their genotoxic effect, sunburn history or history of sunburns in childhood, artificial UV exposure (the amount of UVA occurring in a typical tanning bed session is significantly higher in comparison to the exposure during ordinary outdoor activities or even during sunbathing). The most important host risk factors are the number of melanocytic nevi, familiar history and genetic susceptibility (Rastrelli et al., 2014).

The melanoma development and biology can evolve from different precursor lesions, and can involve different gene mutations and stage of transformation. As example, the BRAF gene is mutated in up to 80% of benign nevi, resulting in limited melanocyte proliferation through the oncogene-mediated activation of cell senescence. Although the oncogenic BRAF itself is not sufficient for melanoma development, generally it is associated with the acquisition of subsequent mutations in key genes, such as TERT or CDKN2A. It is important to understand the biological determinants of melanoma evolution to improve diagnosis and the early recognition of lesions (Leonardi et al., 2018).

The genomic alterations mentioned above lead to the aberrant activation of two main signaling pathways in melanoma: The RAS/RAF/MEK/ERK signaling cascade and the phosphoinositol-3-kinase (PI3K)/AKT pathway. Up to 90% of melanomas exhibit an aberrant MAPK pathway activation and this is a main step in melanoma development, being responsible for cell cycle deregulation and apoptosis inhibition (Wellbrock, Karasarides, and Marais 2004; Leonardi et al., 2018).

Classification of Cutaneous Melanoma

Regarding the clinical and histological features, melanoma can be divided into subtypes. Their relative incidences were: superficial spreading melanoma (50%– 75%), nodular melanoma (15%–35%), lentigo malign melanoma (5%–15%), acral lentiginous melanoma (5%– 10%), desmoplastic melanoma (uncommon), miscellaneous group (Rare) (Bandarchi et al., 2010; Rastrelli et al., 2014).

Superficial Spreading Melanoma (SSM)

It is related to the intermittent exposure to the sun and it is located most often on the back of women's legs and on the men's backs. The classic lesions show variation in pigmentation (tan, brown, gray, black, violaceus, pink and rarely blue or white) and pagetoid spread of melanoma cells in epidermis (Bandarchi et al., 2010; Apalla et al., 2017).

Nodular Melanoma (NMM)

It accounts for 5% of melanomas and most often occurs on the trunk and limbs of patients in the fifth or sixth decade of life; it is more common in males than females. Nodular melanomas are often ulcerated. NMM by definition has no radial growth phase and could be nodular, polypoid, or pedunculated (Rastrelli et al., 2014; Callahan, Flaherty, and Postow 2016).

Lentigo Malign Melanoma (LMM)

It is correlates with long-term sun exposure and increasing age. Clinically, it shows a variety of colors, such as black, brown or brown on a tan background. It has irregular outlines and although the tumor is often relatively large and flat, a focus of invasion may be detected as a papule (Rastrelli et al., 2014).

Acral Lentiginous Melanoma (ALM)

This skin cancer is uncommon, accounting for 5% of melanomas in white people but it is the most common type of melanoma among Asian, Hispanic and African patients. It is mainly located on glabrous skin and adjacent skin of digits, palms and soles. Ulcerate and melanonychia striata may occur (Bandarchi et al., 2010; Rastrelli et al., 2014).

Desmoplastic Melanoma (DM)

Desmoplastic melanoma often occurs in individuals between the age of 60 and 70 years. It rises on the head and neck but it can occur on a variety of cutaneous and mucosal areas. This cancer often shows nerve infiltration and it is characterized by high recurrence rates due to its highly infiltrative growth and frequent perineural invasion. Other rare forms of melanoma have been also described, notably balloon cell melanoma, myxoid melanoma, osteogenic melanoma and rhabdoid melanoma (Rastrelli et al., 2014).

Melanomas display different growth phases: Radial Growth Phase (RGP)- confined microinvasive, which shows some malignant cells present in superficial papillary dermis and Vertical Growth Phase (VGP), which means melanoma has entered the tumorigenic and/or mitogenic phase and can developed metastasis (Callahan, Flaherty, and Postow 2016; Moreno et al., 2018).

Metastatic Melanoma

It is important to determine unknown mechanisms regarding the ability of cancer cells to metastasize, given that tumor metastasis are responsible for most cancer-related deaths (Cazes and Ronai 2016). In the development of malignant melanoma, molecular alterations and protein modifications are responsible for the acquisition of a metastatic phenotype (Gil et al., 2019).

The invasive melanoma cells escape from their primary location and translocate via the circulatory system or body cavities to lymph nodes or distant organs to establish a secondary cancer tissue. In this way advanced melanoma is promoted by a complex interplay between melanoma cells and components of the microenvironment (Gajos-Michniewicz and Czyz 2019).

In order to disseminate into lymph nodes or distant sites, melanoma cells must enter the cardiovascular or lymphatic systems, forming new vessels via angiogenesis, intravasation, and extravasation (Gajos-Michniewicz and Czyz 2019). The metastatic melanoma cells can reprogram their metabolism to survive any stressed conditions and exhibit robust invasiveness capacity due to epigenetic or metabolic changes (Cazes and Ronai 2016).

Besides that, a higher prevalence of comorbidities has been associated with advanced age and more advanced stages of cancer. Increased incidence of comorbidity in advanced metastatic melanoma is emerging as an important factor in patient prognosis, treatment, and survival (Bebe FN, Hu S, Brown TL 2019). About these comorbidities, we can suggest the platelet involvement. Platelets are highly specialized hemostasis effector cells, performing various functions, such as adhesion, aggregation and formation of the hemostatic "buffer." In addition, they participate in additional processes, such as inflammatory processes, vascular and tissue repairment, and modulation of the immune system (Burnstock 2015).

PLATELETS AND CANCER

Platelets were discovered in 1882 by G. Bizzozero. They are small anucleated blood cell fragments released from bone marrow megakaryocytes that circulate in the bloodstream at concentrations of 150 to 350×10^9/L. In cancer, their role was observed over 130 years ago, and trombophilic state, is a frequent finding in cancer patients (Wojtukiewicz et al., 2017; Mezouar et al., 2016).

They are the smallest blood elements (2×10^8/mL) and are characterized by a short turnover time (5 to 7 days). These anucleated blood constituents are surrounded by a phospholipid membrane and composed of glycoproteins, that contribute to the adhesive and aggregative processes. The most important glycoproteins are: Ib-IX-V (GP Ib-IX-V), VI (GP VI), and IIb-IIIa (GP IIb-IIIa, also known as integrin $\alpha IIb\beta 3$) (Wojtukiewicz et al., 2017; Andrews et al., 2001).

Endothelial cell damage or alteration leads to exposure of the subendothelial extracellular matrix (ECM) components that are ligands for platelet adhesion and include various types of collagen, von Willebrand factor (vWF), laminin, vitronectin, proteoglycans, thrombospondin, and fibronectin (Andrews et al., 2001). Platelets are enriched in three types of specific granules (α-granules, dense granules, and lysosomes) that store a diverse array of products, as well as mitochondria and a dense tubular system that facilitates delivery of energy and biochemical messengers that contribute to platelet reactivity (Rendu and Brohard-Bohn 2001).

There is an increased platelet turnover in cancer patients compared to healthy individuals, besides that it's recognized as contribute to metastatic dissemination. Formation of new blood vessels from pre-existing ones is a prerequisite of primary tumor growth, cancer cell intravasation, extravasation, and growth of cancer foci at distant sites. It is widely known that angiogenesis is a rate-limiting process in cancer progression (Wojtukiewicz et al., 2017).

Increased circulating levels of a platelet-specific α-granule protein, β-thromboglobulin, as well as elevated expression of platelet adhesion molecules reflect platelet activation. It was observed significantly elevated levels of β-thromboglobulin in an advanced stage of cancers (Wojtukiewicz et al., 2017; Bambace and Holmes 2011).

Effects of Cancer on Platelets

It has been demonstrated that tumor cells have the ability in vitro to induce platelet activation and aggregation. Several molecular pathways are involved in platelet activation and cancer effects, including the thrombin, ADP, TXA2, metalloproteinases and tissue factor pathways (Umansky et al., 2014; Wachowicz et al., 2002).

- Thrombin: is a multi-functional serine protease that converts fibrinogen to fibrin, activates other proteins in the coagulation cascade, such as coagulation factors V, VIII, XI, and XIII and it is also agonist for platelet activation. Cancer cells directly secrete thrombin, which may activate the coagulation cascade and platelets. In cancer, thrombin signaling was also implicated in tumor cells malignancy and metastasis (Rezer et al., 2016; Bergmeier and Stefanini 2018).
- ADP: is stored in platelet-dense granules and is released upon platelet activation. It interacts with platelet receptors $P2Y_{12}$ and $P2Y_1$, leading to platelet aggregation, shape change and the release of TXA2. ADP is expressed by cancer cells (Atkinson et al., 2006; Burnstock 2015).
- Thromboxane A2: is overexpressed in numerous tumor cell lines and can enhance platelet activation and aggregation (Bergmeier and Stefanini 2018; Andrews et al., 2001).

- Matrix metalloproteinases: are involved in the degradation of the extracellular matrix and have been involved in platelet activation (Gay and Felding-Habermann 2011).
- Tissue factor (TF): is the main activator of the coagulation cascade in physiologic and pathologic conditions. The coagulation cascade is indeed initiated as soon as TF comes into contact with circulating activated FVIIa. This initial contact creates the TF-FVIIa complex, which directly converts FX to activated FXa. Factor Xa complex, with activated FVa and calcium, catalyzes the conversion of prothrombin to thrombin, leading to fibrin formation, platelet activation and thrombus generation. Numerous cancer cell lines express TF and release TF-bearing microparticles released into the circulation with a key role in cancer (Bambace and Holmes 2011).

Effects of Platelets on Cancer

Cancer promotes platelet activation and activated platelets participate in each step of cancer progression (Mezouar et al., 2016; Wojtukiewicz et al., 2017). When activated, platelets are recruited to the tumor microenvironment and can locally regulate tumor cell behavior. During angiogenesis, they come into contact with subendothelial prothrombotic structures. This causes changes in the laminar blood flow, specifically by increasing shear stress, and the VEGF-induced release of vWF by endothelial cells, leading to platelet activation. Activated platelets then release the myriad of angiogenesis-regulating proteins contained in their granules, causing dissemination via the hematogenous circulation. This mechanism seems to represent the major pivot of metastasis (Mezouar et al., 2016; Bambace and Holmes 2011).

- Tumor growth: When activated, platelets recruited to the tumor microenvironment can locally regulate tumor cell behavior and promote proliferation not only through paracrine signaling but also through direct contact with tumor cells. Regarding tumor growth,

platelets not only secrete growth factors but also release toxic cytokines that inhibit tumor growth (Peiris-Pagès et al., 2015).
- Angiogenesis: is defined as the formation of new micro vessels from pre–existing vasculature. It not only prevents both hypoxia and necrosis but also allows the release of proteases and cytokines that contribute to further intravasation, extravasation and dissemination. During angiogenesis, platelets come into contact with subendothelial prothrombotic structures, such as collagen in new vessels with abnormal configuration. This causes changes in the laminar blood flow, specifically by increasing shear stress, and the VEGF-induced release of vWF by endothelial cells, leading to platelet activation (Mezouar et al., 2016; Bambace and Holmes 2011).
- Metastasis: The highly complex process of metastasis includes the detachment of cancer cells from the primary tumor, intravasation into the lumen of blood vessels, intravascular migration, attachment in distant capillaries and finally, extravasation into the surrounding tissue. While the lymphatic spread of cancer cells is observed in human tumors and used as an important prognostic marker for disease progression, dissemination *via* the hematogenous circulation appears to represent the major mechanism of metastasis. During cancer progression, platelets present an abnormal activated state that leads to the formation of platelet-cancer cell aggregates, which ones are crucial in the metastatic process (Mezouar et al.,, 2016; Wojtukiewicz et al.,, 2017).
- Tumor-platelet interaction during the intravascular phase of metastasis: Tumor cells circulating within the vasculature are exposed to high shear stress that can disrupt their membranes. They also have to evade immune surveillance. Because of this, the activation of platelets by tumor cells and the adhesion of platelets to tumor cells are crucial mechanisms allowing the circulation of cancer cells to survive in the bloodstream (Moreno et al., 2018).

Platelets in Melanoma

As mentioned in the previous items, the platelets contain a myriad of bioactive molecules and contribute to inflammation, cancer invasion, and metastasis. Cancer cell derived thrombin is a mediator of this process and facilitates metastasis. When cancer cells leave the primary tumor and enter the blood stream, they activate surrounding platelets via some molecules, inducing microthrombus formation (Figure 1). Mice with defective platelet function are protected from melanoma dissemination to their lungs; however, this protection is mitigated by depletion of natural killer cells. It is known that platelets activate TGFβ and suppress antitumor T lymphocytes. These mechanisms can promote the melanoma progression, as they act directly on cancer cells inducing epithelial-mesenchymal transition, and increasing their metastatic potential (Becker et al., 2017; Umansky et al., 2014).

In a recent study, it was shown that the aspirin used after diagnosis was correlated with longer survival in advanced stages of melanoma. We also observed that cancer was diagnosed in earlier stages with people taking aspirin prior to this diagnosis. In addition, increased levels of circulating platelet-derived microparticles are associated with metastatic cutaneous melanoma (Rachidi et al.,, 2018).

In Rachidi et al., (2019) study, it was observed that thrombocytosis is associated with melanoma distant metastasis and shorter overall survival. Within stage IV, higher platelet counts were correlated with shorter survival in the first year in univariate analysis and after adjusting for age, sex, and treatment.

These data suggest that platelets could serve as a prognostic biomarker in patients with advanced melanoma, and that targeting this cellular entity can have a therapeutic role. It is worth mentioning that mortality from stage IV melanoma, especially prior to the institution of immunotherapy as a first-line treatment, occurred largely in the first year after diagnosis, which probably explains the prognostic significance of platelets mainly in the first year, but not beyond (Rachidi et al.,, 2019).

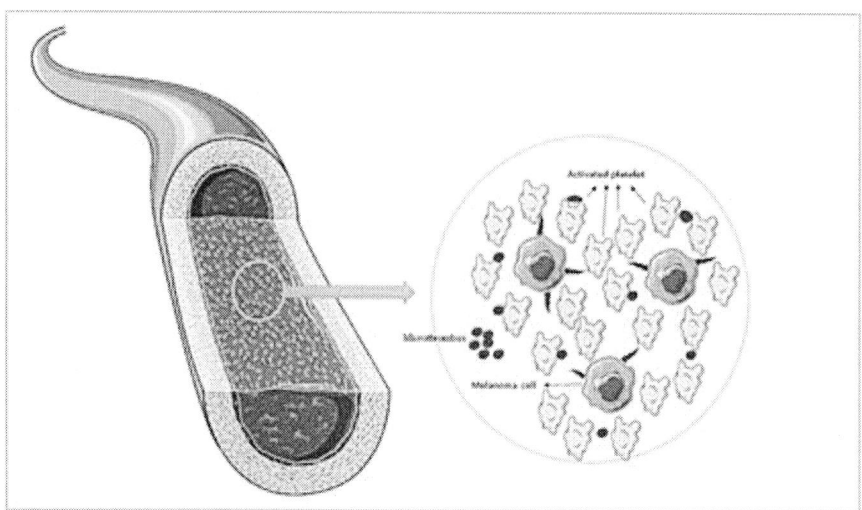

Source: Authors.

Figure 1. Melanoma cells activating the surrounding platelets at blood circulation inducing microthrombus formation and facilitates metastasis.

Another recent study showed that increased platelet-to-lymphocyte ratio correlates with worse outcomes in melanoma, but did not independently evaluate the significance of platelet counts. It is conceivable that patients with more advanced stages are more likely to get blood tests including platelet counts as part of their work up. Indeed, around 70% of our patient population had Breslow thickness > 1 mm, which is an over-representation. This is because blood counts are not part of the standard work up for thin melanomas (Qi et al.,, 2018).

In mouse models' platelets suppress anti-melanoma T cell immunity partly by activating TGFβ (Rachidi et al.,, 2019). Additionally, genetically targeting platelets in mice protected against melanoma dissemination to the lungs. In a separate clinical study, it was observed that intake of aspirin correlated with longer overall survival in melanoma, particularly in stages II–IV (Rachidi et al., 2019; Umansky et al.,, 2014).

Hence targeting platelets is a promising adjuvant therapeutic approach in melanoma and cancer in general. Indeed, clinical trials are currently underway, combining platelet inhibitors such as aspirin and clopidogrel with immunotherapy in metastatic melanoma and head and neck carcinoma.

CONCLUSION

In the present chapter, it was summarized the main mechanisms and role of platelets in cutaneous melanoma disease. It was demonstrated which cancer cells activate platelets and that platelets participate in cancer progression, as well as their role in promoting metastatic melanomas.

It is possible to suggest that targeting platelets are being considered as a promising area for the development of new anticancer and antithrombotic strategies that could improve care and survival for cancer patients, specially melanoma patients.

REFERENCES

Andrews, R. K., Y. Shen, E. E. Gardiner, and M. C. Berndt. 2001. "Platelet Adhesion Receptors and (Patho)Physiological Thrombus Formation." *Histology and Histopathology* 16 (3): 969–80. https://doi.org/10.14670/HH-16.969.

Apalla, Zoe, Dorothée Nashan, Richard B. Weller, and Xavier Castellsagué. 2017. "Skin Cancer: Epidemiology, Disease Burden, Pathophysiology, Diagnosis, and Therapeutic Approaches." *Dermatology and Therapy* 7 (S1): 5–19. https://doi.org/10.1007/ s13555-016-0165-y.

Atkinson, Ben, Karen Dwyer, Keiichi Enjyoji, and Simon C. Robson. 2006. "Ecto-Nucleotidases of the CD39/NTPDase Family Modulate Platelet Activation and Thrombus Formation: Potential as Therapeutic Targets." *Blood Cells, Molecules, and Diseases* 36 (2): 217–22. https://doi.org/10.1016/j.bcmd.2005.12.025.

Bambace, N. M., and C. E. Holmes. 2011. "The Platelet Contribution to Cancer Progression." *Journal of Thrombosis and Haemostasis* 9 (2): 237–49. https://doi.org/10.1111/j.1538-7836.2010.04131.x.

Bandarchi, Bizhan, Linglei Ma, Roya Navab, Arun Seth, and Golnar Rasty. 2010. "From Melanocyte to Metastatic Malignant Melanoma." *Dermatology Research and Practice* 2010 (1): 1–8. https://doi.org/10.1155/2010/583748.

Bebe FN, Hu S, Brown TL, Tulp OL. 2019. *Role, Extent, and Impact of Comorbidity on Prognosis and Survival in Advanced Metastatic Melanoma: A Review* 12 (1): 16–23. https://doi.org/10.1 4670/HH-16.969.

Becker, Katrin Anne, Nadine Beckmann, Constantin Adams, Gabriele Hessler, Melanie Kramer, Erich Gulbins, and Alexander Carpinteiro. 2017. "Melanoma Cell Metastasis via P-Selectin-Mediated Activation of Acid Sphingomyelinase in Platelets." *Clinical & Experimental Metastasis* 34 (1): 25–35. https://doi.org/10.1007/s10585-016-9826-6.

Bergmeier, Wolfgang, and Lucia Stefanini. 2018. "Platelets at the Vascular Interface." *Research and Practice in Thrombosis and Haemostasis* 2 (1): 27–33. https://doi.org/10.1002/rth2.12061.

Burnstock, Geoffrey. 2015. "Blood Cells: An Historical Account of the Roles of Purinergic Signalling." *Purinergic Signalling* 11 (4): 411–34. https://doi.org/10.1007/s11302-015-9462-7.

Callahan, Margaret K, Catherine R Flaherty, and Michael A Postow. 2016. Melanoma. Edited by Howard L. Kaufman and Janice M. Mehnert. Vol. 167. *Cancer Treatment and Research.* Cham: Springer International Publishing. https://doi.org/10.1007/978-3-319-22539-5.

Cazes, Alex, and Ze'ev A. Ronai. 2016. "Metabolism in Melanoma Metastasis." *Pigment Cell & Melanoma Research* 29 (2): 118–19. https://doi.org/10.1111/pcmr.12440.

d'Ischia, Marco, Kazumasa Wakamatsu, Fabio Cicoira, Eduardo Di Mauro, Josè Carlos Garcia-Borron, Stephane Commo, Ismael Galván, et al., 2015. "Melanins and Melanogenesis: From Pigment Cells to Human Health and Technological Applications." *Pigment Cell & Melanoma Research* 28 (5): 520–44. https://doi.org/10.1111/pcmr. 12393.

Gajos-Michniewicz, Anna, and Malgorzata Czyz. 2019. "Role of MiRNAs in Melanoma Metastasis." *Cancers* 11 (3): 326. https://doi.org/10.3390/cancers11030326.

Garbe, C., T. K. Eigentler, U. Keilholz, A. Hauschild, and J. M. Kirkwood. 2011. "Systematic Review of Medical Treatment in Melanoma: Current Status and Future Prospects." *The Oncologist* 16 (1): 5–24. https://doi.org/10.1634/theoncologist.2010-0190.

Gay, Laurie J., and Brunhilde Felding-Habermann. 2011. "Contribution of Platelets to Tumour Metastasis." *Nature Reviews Cancer* 11 (2): 123–34. https://doi.org/10.1038/nrc3004.

Gil, Jeovanis, Lazaro Hiram Betancourt, Indira Pla, Aniel Sanchez, Roger Appelqvist, Tasso Miliotis, Magdalena Kuras, et al., 2019. "Clinical Protein Science in Translational Medicine Targeting Malignant Melanoma." *Cell Biology and Toxicology* in press (March). https://doi.org/10.1007/s10565-019-09468-6.

Leonardi, Giulia, Luca Falzone, Rossella Salemi, Antonino Zanghï¿½, Demetrios Spandidos, James Mccubrey, Saverio Candido, and Massimo Libra. 2018. "Cutaneous Melanoma: From Pathogenesis to Therapy (Review)." *International Journal of Oncology* 52 (4): 1071–80. https://doi.org/10.3892/ijo.2018.4287.

Mezouar, Soraya, Corinne Frère, Roxane Darbousset, Diane Mege, Lydie Crescence, Françoise Dignat-George, Laurence Panicot-Dubois, and Christophe Dubois. 2016. "Role of Platelets in Cancer and Cancer-Associated Thrombosis: Experimental and Clinical Evidences." *Thrombosis Research* 139 (March): 65–76. https://doi.org/10.1016/j.thromres.2016.01.006.

Moreno, Marcelo, Bruna Conte, Eduardo Menegat, Yixuan James Zheng, Ricardo Moreno Traspas, Susana Ortiz-Urda, Marcela Valko-Rokytovská, et al., 2018. "Possibilities for the Therapy of Melanoma: Current Knowledge and Future Directions." *Human Skin Cancers - Pathways, Mechanisms, Targets and Treatments* 3 (1): 3–24. https://doi.org/10.1007/s12307-018-0206-4.

Peiris-Pagès, Maria, Ubaldo E. Martinez-Outschoorn, Federica Sotgia, and Michael P. Lisanti. 2015. "Metastasis and Oxidative Stress: Are Antioxidants a Metabolic Driver of Progression?" *Cell Metabolism* 22 (6): 956–58. https://doi.org/10.1016/j.cmet.2015.11.008.

Qi, Yalong, Yong Zhang, Xiaomin Fu, Axiang Wang, Yonghao Yang, Yiman Shang, and Quanli Gao. 2018. "Platelet-to-Lymphocyte Ratio in Peripheral Blood: A Novel Independent Prognostic Factor in Patients with Melanoma." *International Immunopharmacology* 56 (August 2017): 143–47. https://doi.org/10.1016/j.intimp.2018.01.019.

Rachidi, Saleh, Maneet Kaur, Tim Lautenschlaeger, and Zihai Li. 2019. "Platelet Count Correlates with Stage and Predicts Survival in Melanoma." *Platelets* 00 (00): 1–5. https://doi.org/10.1080/09537104.2019.1572879.

Rachidi, Saleh, Kristin Wallace, Hong Li, Tim Lautenschlaeger, and Zihai Li. 2018. "Postdiagnosis Aspirin Use and Overall Survival in Patients with Melanoma." *Journal of the American Academy of Dermatology* 78 (5): 949–956.e1. https://doi.org/10.1016/j.jaad.2017.12.076.

Rastrelli, Marco, Saveria Tropea, Carlo Riccardo Rossi, and Mauro Alaibac. 2014. "Melanoma: Epidemiology, Risk Factors, Pathogenesis, Diagnosis and Classification." *In Vivo* 1012: 1005–11. https://doi.org/0258-851X/2014.

Rendu, Francine, and Brigitte Brohard-Bohn. 2001. "The Platelet Release Reaction: Granules' Constituents, Secretion and Functions." *Platelets* 12 (5): 261–73. https://doi.org/10.1080/09537100120068170.

Rezer, João Felipe P., Viviane C.G. Souza, Maria Luiza P. Thorstenberg, Jader B. Ruchel, Tatiana M.D. Bertoldo, Daniela Zanini, Karine L. Silveira, et al., 2016. "Effect of Antiretroviral Therapy in Thromboregulation through the Hydrolysis of Adenine Nucleotides in Platelets of HIV Patients." *Biomedicine & Pharmacotherapy* 79 (April): 321–28. https://doi.org/10.1016/j.biopha.2016.02.008.

Saida, Toshiaki. 2019. "Histogenesis of Cutaneous Malignant Melanoma: The Vast Majority Do Not Develop from Melanocytic Nevus but Arise de Novo as Melanoma in Situ." *The Journal of Dermatology* 46 (2): 80–94. https://doi.org/10.1111/1346-8138.14737.

Tsao, Hensin, Lynda Chin, Levi A. Garraway, and David E. Fisher. 2012. "Melanoma: From Mutations to Medicine." *Genes and Development* 26 (11): 1131–55. https://doi.org/10.1101/gad.191999.112.

Umansky, Viktor, Ivan Shevchenko, Alexandr V. Bazhin, and Jochen Utikal. 2014. "Human Melanoma Cell Lines Differ in Their Capacity to Release ADP and Aggregate Platelets." *Cancer Immunology, Immunotherapy* 63 (10): 1073–80. https://doi.org/10.1111/j.1365-2141.1994.tb06736.x.

Wachowicz, B., B. Olas, H. M. Zbikowska, and Andrzej Buczyński. 2002. "Generation of Reactive Oxygen Species in Blood Platelets." *Platelets* 13 (3): 175–82. https://doi.org/10.1080/09533710022149395.

WallaCeH. Clark, Jr., Lynn From, Evelina A. Bernardino, and Martin C. Mihm Departments. 1969. "The Histogenesis and Biologic Behavior of Primary Human Malignant Melanomas of the Skin." *Cancer RESEARCH* 29 (1): 705–26. http://journal.hsforum.com/index.php/HSF/article/view/242.

Wellbrock, Claudia, Maria Karasarides, and Richard Marais. 2004. "The RAF Proteins Take Centre Stage." *Nature Reviews Molecular Cell Biology* 5 (11): 875–85. https://doi.org/10.1038/nrm1498.

Wojtukiewicz, Marek Z., Ewa Sierko, Dominika Hempel, Stephanie C. Tucker, and Kenneth V. Honn. 2017. "Platelets and Cancer Angiogenesis Nexus." *Cancer and Metastasis Reviews* 36 (2): 249–62. https://doi.org/10.1007/s10555-017-9673-1.

In: Platelets
Editor: René Langelier

ISBN: 978-1-53616-592-0
© 2019 Nova Science Publishers, Inc.

Chapter 3

AUTOLOGOUS PLATELET-RICH PLASMA (PRP): A TREATMENT FOR DIFFICULT-TO-HEAL VENOUS LEG ULCERS

Valeria G. Mateeva*, MD, PhD,
Doncho N. Etugov, MD, PhD
and Grisha S. Mateev, MD, PhD
Department of Dermatology and Venereology, Medical University - Sofia, Sofia, Bulgaria

ABSTRACT

Venous leg ulcers (VLUs) are a common medical problem in everyday practice. The condition is encountered in around 5% of the elderly above 65 years and in 1.5% of the general population. The management of VLUs is interdisciplinary and presents a significant economic burden.

---

* Corresponding Author's E-mail: vali_mateeva@hotmail.com.

Furthermore, the importance of VLUs is increasing due to the worldwide demographic tendency for aging of the population.

Conventional methods for treatment of VLUs include: compressive therapy, local wound care such as mechanical debridement and local antibiotics in case of infection; systemic medication and surgery. Often these methods require prolonged treatment and lack in efficacy.

The fundamental pathophysiological process in VLUs is inflammation of the venous wall due to increased hydrostatic pressure in the vascular system of the lower extremities.

The induced inflammatory response consists of interplay between inflammatory cells such as leukocytes, macrophages and monocytes, T-lymphocytes, and inflammatory molecules such as mediators, chemokines, growth factors, etc.

Growth factors derived from the thrombocytes modify the microenvironment of the wound and stimulate the capillary angiogenesis, fibroblastic migration and proliferation, collagen synthesis and reepitelisation. The most important thrombocytic growth factors are: platelet-derived growth factor (PDGF), platelet-derived angiogenesis factor (PDAF), platelet-derived epidermal growth factor (PDEGF) and platelet factor 4.

The role of the platelet-rich-plasma (PRP) in stimulating the healing process in difficult-to-heal ulcers has been investigated in the past 25 years. It is suggested that PRP is capable to transform the difficult-to-heal skin ulcer with low metabolic activity into a healing ulcer with increased capacity for tissue regeneration.

Keywords: Difficult-to-heal venous leg ulcers, autologous platelet-rich plasma

INTRODUCTION

Venous leg ulcers (VLUs) are a common medical problem in everyday practice. The condition is encountered in around 5% of the elderly above 65 years [1] and in 1.5% of the general population [2]. The management of VLUs is interdisciplinary and presents a significant economic burden. Furthermore, the importance of VLUs is increasing due to the worldwide demographic tendency for aging.

VLUs share the common features of all skin ulcerations and are generally characterized by loss of tissue, which involves the epidermis, the dermis, and, occasionally, the hypodermis and the fascia of the underlying muscles. They heal with formation of a scar [3]. Besides the VLUs, other causes for skin ulcers include: infections, trauma, neurological, neoplastic and metabolic diseases, or a combination of two or more of those conditions [4]. High risk patients to develop difficult-to-treat ulcers include diabetic patients and cancer patients who receive immunosuppressive cytostatic therapy or radiotherapy.

The chronic difficult-to-heal VLUs have an important social impact, since they may lead to a significant decrease of the productivity and the quality of life of the affected patients [5-9].

PATHOPHYSIOLOGY AND PRINCIPLES OF TISSUE REGENERATION IN VLUS

The fundamental pathophysiological process in VLUs is inflammation of the venous wall due to increased hydrostatic pressure in the vascular system of the lower extremities. This may be complicated by a disorder in the normal process of tissue regeneration such as decrease in the oxygen perfusion and infection [10-12]. In diabetic patients, the difficult-to-heal ulcers cause more complications, since they develop on a pathological background of decreased tissue reactivity, reduced vascularisation and impaired local and general immunity.

A key step in the process of tissue regeneration in chronic VLUs is the transformation of the connective tissue into an area of active cell proliferation, neovascularisation, biosynthesis of new collagen fibers and intensive intercellular exchange. This step is accomplished under the action of growth factors and other signal molecules. The process typically evolves into three stages: inflammatory stage, proliferation stage and tissue remodeling. The last stage is followed by formation of granulation tissue and reepithelisation [13]. All skin ulcerations, including the VLUs, heal with the formation of a residual scar.

The regulation of the process of healing of the skin ulcers includes interplay of thrombocytes, monocytes, fibroblasts, endothelial cells, epidermal cells, serum enzymes, locally active growth factors, and factors of the local cell microenvironment [13-15]. The current findings show that the thrombocytes and marcophages are the dominant cell types in this process [16].

The growth factors derived from the thrombocytes act in the early stages of inflammation and cicatrisation. The main thrombocytic growth factors are: platelet-derived growth factor (PDGF), platelet-derived angiogenesis factor (PDAF), platelet-derived epidermal growth factor (PDEGF) and platelet factor 4. These factors are known to stimulate the capillary neoangiogenesis, the fibroblast migration and proliferation, the collagen synthesis and the reepithelisation.

Specifically, the PDGF is a powerful firoblastic mitogen factor and chemoattractant [17-20]. PDAF leads to the formation of novel capillary network [16]. The angiogenesis is mediated by the pericytes, which are sensitive to the PDGF and the vascular endothelial growth factor (VEGF). Platelet factor 4 acts as a neutrophilic chemoattractant.

The fibroblasts proliferate and synthesize collagen 1 and collagen 3, reticulin, elastin and hyaluronic acid, which play an important role in the tissue regeneration. Through autocrine secretion of TGF β and PDGF the fibroblasts stimulate their own proliferation.

The last stage of regeneration is reepithelisation, which is characterized by the proliferation of keratinocytes under the stimulation of proinflammatory cytokines, fibroblastic growth factors (epithelial growth factor, EGF; keratocyte growth factor, KGF), and factors, derived from the keratinocytes [21].

The difficult-to-heal skin ulcers are marked by a deficiency of growth factors. Based on this conception, new treatment methods arise. They aim at facilitation of the regeneration process by a local administration of growth factors [22-24].

PLATELET-RICH-PLASMA (PRP) IN VLUS

The conventional methods for treatment of the difficult-to-heal ulcers are based on the restoration of the physiological characteristics of the tissue in order to promote healing. Those methods involve: debridement of necrotic tissues, evacuation of the exudate, application of protective dressings, local and systemic antibiotics against infections, etc. The therapeutic results are usually delayed and a prolonged treatment period is often required (an average of 198 weeks [25].

Nowadays, new treatment modalities for VLUs emerge: hyperbaric oxygenation, electrical stimulation, application of growth factors, etc. The growth factors are a promising new therapeutic option for the condition, since they are capable to induce tissue proliferatio [26, 27].

The role of the platelet-rich-plasma (PRP) in stimulating the healing process in difficult-to-heal ulcers has been investigated in the past 25 years. It is suggested that PRP is capable to transform the difficult-to-heal skin ulcer with low mitogenic activity into a healing ulcer with increased capacity for tissue regeneration [28].

The platelet-rich plasma (PRP) contains growth factors from the thrombocytes. The PRP is injected at the site of the VLU or applied in the form of a gel and it accelerates the process of cicatrisation.

PRP is indicated in all forms of chronic difficult-to-heal ulcers with exception of the cases of thrombocytopenia, disturbance in the normal function of the platelets, anemia, skin carcinoma, or active infection, which requires antibiotic therapy [29-31].

The autologous plasma stimulates the migration of the peripheral immune cells, such as monocytes and macrophages.

The monocytes play a special role, since they do not only participate in the immune defense against bacteria, but also stimulate the fibroblastic proliferation through the secretion of proinflammatory cytokines. The circulating monocytes in the peripheral blood transform into tissue macrophages at the site of damage [16].

Similar action is characteristic for the histiocytes through IGF1, TGF β, TNF alpha and PDGF. The macrophages have a leading role in the regeneration in the 24 hours following the tissue damage [32]. As a result of the local hypoxia at the site of damage, the macrophages secrete factors, that stimulate the angiogenesis [33-34]. Another type macrophage-derived growth factors lead to an activation and proliferation of the fibroblasts.

METHODOLOGY

Preparation of the VLU

All infections should be eradicated prior to the application of the PRP. Afterwards, a debridement should be performed in order to eliminate the necrotic tissue. This procedure should be repeated before each application of the PRP, if indicated [12].

Preparation and Application of the PRP

A large number of methods for preparation of PRP are described in the literature. Possible methods are presented below:

PRP for Injection at the Site of the Wound
After applying topical anesthesia on the treated area (lidocaine, prilocaine) under occlusion for 50 min., 10 ml blood is extracted from the patient and centrifuged at 3100/min for 4 min. A standardized kit is used for the procedure. The separated platelet-rich-plasma is aspirated with a 20G needle in a 20 cc syringe and is injected in the bottom and edges of the ulcer and 3-5 cm in surrounding skin. The surface of the ulcer is divided into $1 cm^2$ squares and 0,2 cc PRP is injected in the middle of each of them. Occlusive dressing with sterile gauze is applied for 5 days. After removal of the dressing, the ulcer is being cleaned with saline and covered with silver sulfadiazine 1% cream every second day [35].

Other authors suggest injecting 1ml PRP in every 1cm² of the ulcer. Half of the quantity should be injected deeply, while the remaining half may be applied at the surface of the ulcer under occlusive dressing [36].

Gel, Containing Platelet-Derived Wound Healing Factors (PDWHF)

The platelet-rich gel is prepared from 20-60ml blood (for small-sized ulcers) to 240ml blood (for large VLUs). The blood is drawn directly before the preparation and use of the gel [37].

Knighton et al. propose the following preparation method: 60ml blood is obtained from the patient in a tube with 6ml anticoagulant and citrate dextrose. The whole blood is centrifuged at 135 x g for 20 min. at 4^0C. to separate and eliminate the erythrocytes and the leukocytes. The so obtained PRP is further centrifuged at 750 x g for 10 min. at 4^0C to separate the thrombocytes from the plasma. The plasma is then eliminated. The thrombocytes are washed with buffer and then a suspension of a standardized concentration of 10^9 thrombocytes /ml is produced. Thrombin 1U/ml is added and the produced supernatant contains PDWHF. Additionally 10 ml PDWHF are added to 1g microcrystalline collagen to produce the stable sterile gel.

The gel is applied in thin layer on the ulcer and is covered with a vaseline impregnated sterile gauze. After 12 hours the gauze is removed and the wound is washed with buffer. Silver sulfadiazine 1% cream is applied for 12 hours. The cream is afterwards washed with water [12].

CLINICAL RESULTS

Systematic review of 18 clinical studies on the effectiveness of PRP on the healing process in chronic leg ulcers, conducted in the period between 1996 and 2006, proves that PRP could be considered a method of choice in these types of ulcers [37].

A study of Knighton et al. on 49 patients with difficult-to-treat chronic ulcers of various origins (venous stasis, arterial insufficiency, diabetes, posttransplatational, decubital wounds) demonstrates that a single application of platelet-derived wound healing factors accelerates the formation of granulation tissue and facilitates the reepitelisation of the ulcer [16].

Crovetti et al. [38] confirm the beneficial effect of the administration of platelet-rich gel on the process of tissue reconstruction in 24 patients with chronic non-healing ulcers. In 9 of them the authors report complete wound healing after 10 applications of the gel, and in other 9 - a partial response to the treatment.

Mazucco et al. [39] report similar results by comparing 17 patients with chronic non-healing ulcers (postsurgical dehisced sternal wounds and necrotic skin ulcers) treated with platelet-rich gel with 14 controls treated with conventional methods. Results show that the healing time in the investigated group is shortened by approximately 50%.

A number of subsequent studies confirm the positive effect of the platelet rich media on the healing of chronic skin ulcers [40-44].

A recent controlled study by Etugov et al. on 23 patients with VLUs shows significant reduction in the size of the ulcer both in the group treated with a single application of PRP and in the control group (CG) treated with conventional methods. The reduction in the size of the VLU is more pronounced in the group, treated with PRP compared to the CG [35].

Nevertheless, despite the above mentioned existing data, which support the use of PRP in the treatment of chronic ulcers, a metaanalysis of 9 randomized controlled trials (RCTs) with 325 patients, fails to demonstrate a statistically significant difference between PRP and placebo in influencing the healing process [45].

The difference in the reported outcomes may be due to the different design of the trials and the resulting difficulty in comparing the data.

Conclusion

The application of PRP in difficult-to-treat VLUs may be a promising new method for the therapy of this problem. PRP improves all aspects of the tissue regeneration and reduces the duration of the treatment. Nevertheless, further research in the area is needed to evaluate the therapeutic significance of the method and to prove its potential superiority to conventional treatments in larger cohorts of patients.

REFERENCES

[1] Guest JF, Ruiz FJ, Mihai A, Lehman A. 2005. "Cost effectiveness of using carboxymethylcellulose dressing compared with gauze in the management of exuding venous leg ulcers in Germany and the USA." *Curr. Med. Res. Opin.* 21:81-92.

[2] Partsch H. 2008. "Intermittent pneumatic compression in immobile patients." *Int. Wound. J.* 5:389-397.

[3] McGrath J., Breathnach S. 1998. "Wound Healing." In: *Textbook of Dermatology. 6th ed.*, edited by Champion RH., Burton JL., Burns AD., et al. Oxford: Blackwell.

[4] Liu C, Zhang H. 2014. "Advances in the research of promoting healing of chronic wound with platelet-rich plasma." *Zhonghua Shao Shang Za Zhi* 30(5):433-6.

[5] Harrington C, Zagari MJ, Corea J, et al. 2000. "A cost analysis of diabetic lower-extremity ulcers." *Diabetes Care* 23:1333-1338.

[6] Harding KG, Morris HL, Patel GK. 2002. "Healing chronic wounds." *Br. Med. J.* 324:160-163.

[7] Albert S. 2002. "Cost-effective management of recalcitrant diabetic foot ulcers." *Clin. Pediatr. Med. Surg.* 19:483-491.

[8] Vileikyte L. 2001. "Diabetic foot ulcers. A quality of life issue." *Diabetes Metab. Res. Rev.* 7:246-249.

[9] Phillips T, Stanton B, Provan A, et al. 1994. "A Study of the impact of leg ulcers on quality of life: financial, social and psychological implications." *J. Am. Acad. Dermatol.* 31:49-53.

[10] Eaglstein WH, Falanga V. 1997. "Chronic wounds." *Surg. Clin. North. Am.* 77:689–97.

[11] Stadelmann WK, Digenis VD, et al. 1998. *"Impediments to wound healing. Am. J. Surg."* 176:39–47.

[12] Knighton DR, Fiegel VD, et al. 1986. "Classification and treatment of chronic nonhealing wounds." *Ann. Surg.* 204:322–330.

[13] Goss JR. 1992. "Regeneration versus repair." In: *Wound healing*, edited by Cohen IK, Diegelmann RF, Lindblad WJ, 40–62. Philadelphia: Saunders.

[14] Hunt TK., Van Winkle WJr. 1976. "Fundamentals of wound management in surgery." In *Wound Healing: Disorders of Repair*. South Plainfield, NJ: Chirugicom.

[15] Rothe M, Falanga V. 1989. "Growth factors." *Arch. Dermatol.* 125:1390–1398.

[16] Knighton DR., Hunt TK., Thakral KK., Goodson WH. 1982. "Role of platelets and fibrin in the healing sequence: an in vivo study of angiogenesis and collagen synthesis." *Ann. Surg.* 196:379-388.

[17] Ross R, Glomset J, Kariya B, et al. 1974. "A platelet-dependent serum factor that stimulates the proliferation of arterial smooth muscle cells in vitro." *Proc. Natl. Acad. Sci. USA* 71:1207-1210.

[18] Antoniodes HN, Scher CD, Stiles CD. 1979. "Purification of human platelet-derived growth factor." *Proc. Natl. Acad. Sci. USA* 76:1809-1813.

[19] Grotendorst GR, Martin GR, Pencev D, et al. 1985. "Stimulation of granulation tissue formation by platelet-derived growth factor in normal and diabetic rats." *J. Clin. Invest.* 76:2323-2329.

[20] Shimokado K, Raines EW, Madtes DK, et al. 1985. "A significant part of macrophage-derived growth factor consists of at least two forms of PDGF." *Cell* 43:277-286.

[21] Bioulac. 2012. "Le plasma autologue et les pathologies cutanees." *J. Med. Esth. Et. Chir. Derm.* 39:1-3.

[22] Ksander GA, Chu GH, McMullin H, et al. 1990. "TGFs-beta 1 and beta 2 enhance connective tissue formation in animal models of dermal wound healing by secondary intent." *Ann. NY Acad. Sci.* 593:135–147.

[23] Uhl E, Barker JH, Bondar I, et al. 1993. "Basic fibroblast growth factor accelerates wound healing in chronically ischaemic tissue." *Br. J. Surg.* 80(8):977–980.

[24] Xia YP, Zhao Y, Marcus J, et al. 1999. "Effects of keratinocyte growth factor—on wound healing in a ischaemia-impaired rabbit ear model and on scar formation." *J. Pathol.* 188:431–438.

[25] Knighton D, Fiegel V, Lorinda L, et al. 1986. "Growth Factor Stimulated Clinical Wound Repair." *Classification and treatment of chronic nonhealing wounds* 204(3): 322-329.

[26] Mason J, O'Keeffe C, Hutchinson A, et al. 1999. "A systematic review of foot ulcer in patients with type 2 diabetes mellitus: II. Treatment." *Diab. Med.* 16:889-909.

[27] Houghton PE, Kincaid CB, Lovell M, et al. 2003. "Effect of electrical stimulation on chronic leg ulcer size and appearance." *Phys. Ther.* 83:17-28.

[28] Roubelakis MG, Trohatou O, Roubelakis A, Mili E, Kalaitzopoulos I, Papazoglou G, Pappa KI, Anagnou NP. 2014. "Platelet-rich plasma (PRP) promotes fetal mesenchymal stem/stromal cell migration and wound healing process." *Stem Cell Rev.* 10(3):417-28.

[29] Weibrich G, Kleis W, Hafner G. 2002. "Growth factor levels in the platelet rich plasma produced by 2 different methods: Curasan-type PRP kit versus PCCS PRP system." *Oral Maxillofac. Implants.* 17:184-190.

[30] Marx R, Carlson ER, Eichstaedt RM. 1998. "Platelet rich plasma: Growth factor enhancement for bone grafts." *Oral Surg* 85:638.

[31] Clark, Richard. 1996. *The Molecular and Cellular Biology of Wound Repair, 2nd ed.* New York, London: Penum.

[32] Hunt TK, Knighton DR, Thakral KK, et al. 1984. "Studies on inflammation and wound healing: angiogenesis and collagen synthesis stimulated in vivo by resident and activated wound macrophages." *Surgery* 96:48-54.

[33] Knighton DR, Hunt TK, Scheuenstuhl H, et al. 1983. "Oxygen tension regulates the expression of angiogenesis factor by macrophages." *Science* 221:1283-1285.

[34] Knighton DR, Silver IA, Hunt TK. 1981. "Regulation of wound-healing angiogenesis: effect of oxygen gradients and inspired oxygen concentration." *Surgery* 90:262-270.

[35] Etugov D, Mateeva V, Mateev G. 2018. "Autologous platelet-rich plasma for treatment of venous leg ulcers: a prospective controlled study." *J. Biol. Regul. Homeost. Agents* 32(3):593-597.

[36] Yotsu RR, Hagiwara S, Okochi H., et al. 2015. "Case series of patients with chronic foot ulcers treated with autologous platelet-rich plasma." *J. Dermatol.* 42(3):288-295.

[37] Villela DL., Santos VL. 2010. "Evidence on the use of platelet-rich plasma for diabetic ulcer: a systematic review." *Growth Factors* 28(2):111-116.

[38] Crovetti G, Martinelli G, Issi M, Barone M, Guizzardi M, Campanati B, Moroni M, Carabelli A. 2004. "Platelet gel for healing cutaneous chronic wounds." *Transfus. Apher. Sci.* 30(2):145-51.

[39] Mazzucco L, Medici D, Serra M, Panizza R, Rivara G, Orecchia S, Libener R, Cattana E, Levis A, Betta PG, Borzini P. 2004. "The use of autologous platelet gel to treat difficult-to-heal wounds: a pilot study." *Transfusion* 44(7):1013-8.

[40] Yilmaz S, Aksoy E, Doganci S, Yalcinkaya A, Diken AI, Cagli K. 2015. "Autologous platelet-rich plasma in treatment of chronic venous leg ulcers: A prospective case series." *Vascular* 23(6):580-5.

[41] Salazar-Álvarez AE, Riera-del-Moral LF, García-Arranz M, Alvarez-García J, Concepción-Rodriguez NA, Riera-de-Cubas L. 2014. "Use of platelet-rich plasma in the healing of chronic ulcers of the lower extremity." *Actas Dermosifiliogr.* 105(6):597-604.

[42] Salcido RS. 2013. "Autologous platelet-rich plasma in chronic wounds." *Adv. Skin Wound Care* 26(6):248.

[43] Dionyssiou D, Demiri E, Foroglou P, Cheva A, Saratzis N, Aivazidis C, Karkavelas G. 2013. "The effectiveness of intralesional injection of platelet-rich plasma in accelerating the healing of chronic ulcers: an experimental and clinical study." *Int. Wound. J.* 10(4):397-406.

[44] Alavi A, Sibbald RG, Phillips TJ, Miller OF, Margolis DJ, Marston W, Woo K, Romanelli M, Kirsner RS. 2016. "What's new: Management of venous leg ulcers: Treating venous leg ulcers." *J. Am. Acad. Dermatol.* 74(4):643-64.

[45] Martinez-Zapata MJ, Martí-Carvajal AJ, Solà I, Expósito JA, Bolíbar I, Rodríguez L, Garcia J, Zaror C. 2016. "Autologous platelet-rich plasma for treating chronic wounds. *Cochrane Database Syst. Rev.*" May 25;(5):CD006899. Accessed December 28, 2018.

In: Platelets
Editor: René Langelier
ISBN: 978-1-53616-592-0
© 2019 Nova Science Publishers, Inc.

Chapter 4

THROMBOCYTOPENIA: EMPHASIS ON ETIOLOGY AND THERAPEUTICS

R. Vani and M. Manasa
Department of Biotechnology, JAIN (Deemed-to-be University),
Bengaluru, Karnataka, India

ABSTRACT

Platelets are anucleate cells, of 1-2 microns and are the second most abundant cells in blood. Their concentration in blood ranges between 150 - 450 x 10^9/L in an average adult human. Megakaryocytes are the precursors of platelets in bone marrow, from which proplatelets are released into the bloodstream due to shear stress. Humans produce ~10^{11} platelets per day, and with a life span of 7 – 10 days in circulation. Platelets have a major role in repairing the damaged endothelium of blood vessels. They arrest bleeding at the site of vascular injury through a process involving platelet adhesion, activation, secretion, and aggregation, subsequently leading to the formation of a hemostatic plug.

Thrombocytopenia is a platelet disorder caused due to the reduction in platelet number (< 150 x 10^9 cells/L). It is further classified based on the platelet count, as mild (> 70 x 10^9 cells/L), moderate (20 - 70 x 10^9 cells/L) and severe (< 20 x 10^9 cells/L). Severe thrombocytopenia can be life-threatening. Thrombocytopenia can be attributed to factors such as (a) increased platelet destruction, (b) reduced platelet production and (c) abnormal platelet distribution/splenic pooling, or a combination of these factors.

Thrombocytopenia can be additionally classified based on pathogenesis as:

1. Pseudo thrombocytopenia is due to *ex vivo* agglutination of platelets. This condition is observed when EDTA is used as an anticoagulant. However, it is necessary to confirm pseudo thrombocytopenia by manual examination of a blood smear to avoid unnecessary tests and treatment.
2. Thrombocytopenia due to reduced platelet production.
    (a) Inherited platelet disorders are caused due to genetic defects affecting the production of platelets, their functions, and morphology.
    (b) Acquired platelet disorders can be immune-mediated, drug-induced, pregnancy-related or due to nutritional deficiencies.
3. Thrombocytopenia due to elevated platelet destruction can be drug-induced, immune-mediated, or related to artificial surfaces (like materials used during hemodialysis or cardiopulmonary bypass).
4. Thrombocytopenia due to abnormal platelet distribution can be caused by hypersplenism, hypothermia or transfusions.
5. Thrombocytopenia due to other causes include cyclic thrombocytopenia and acquired pure megakaryocytic thrombocytopenia.

Various causative factors form the basis of the strategies employed in thrombocytopenia treatments. Essentially, withdrawal of the drug causing thrombocytopenia is an important therapeutic measure in drug-induced thrombocytopenia. Treatment for immune thrombocytopenia includes glucocorticoids, intravenous immunoglobulins, splenectomy, anti-(Rh)D, thrombopoietin receptor agonists, drugs like rituximab, romiplostim, eltrombopag, azathioprine, cyclophosphamide, cyclosporine, vinca alkaloids, etc. However, treatment of the underlying disease, apart from treating thrombocytopenia, is recommended for patients with splenomegaly and viral infections.

This chapter focuses on the causes, diagnosis, and prognosis of various types of thrombocytopenia. It also gives an outline on future prospects of using antioxidants for the treatment of a few thrombocytopenic conditions.

Keywords: platelet disorders, thrombocytopenia, antioxidants, therapeutics

INTRODUCTION

Platelets (thrombocytes) are anucleate cells, of 1-2 microns and are the second most abundant cells in blood. Megakaryocytes (large, polyploid marrow cells) are the precursors of platelets, from which proplatelets are released into the bloodstream due to shear stress. This process is known as megakaryopoiesis. Humans produce approximately 10^{11} platelets per day. One-third of these platelets are sequestered in spleen, while the remaining two-thirds circulate with a life span of 7 – 10 days. The platelet concentration in blood ranges between 150-450 x 10^9/L in an average adult human, only a part of which is consumed for hemostasis and wound healing process. Other platelets circulate until they are senescent and are cleared by the macrophages of the reticulo-endothelial system. (Kasper et al., 2005, Kaushansky et al., 2016).

Platelets play a major role in repairing the damaged endothelium of blood vessels. Their shape and small size enable them to circulate near the wall of the blood vessels, placing them in the optimum location required to constantly survey the integrity of the vasculature. They arrest bleeding at the site of vascular injury through a process involving platelet adhesion, activation, secretion, and aggregation, subsequently leading to the formation of a hemostatic plug. Other functions of platelets include inflammation, host defense and tumor growth (Johns, 2004; Harrison, 2005).

Abnormalities in platelets (numbers and function) lead to various bleeding disorders since platelets form the first line of defense against bleeding (Kasper et al., 2005). Thrombocytopenia is characterized by reduced platelet numbers (< 150 x 10^9/L). The three stages of thrombocytopenia include (i) mild (platelet concentration between 100-150 x 10^9/L), (ii) moderate (50 - 100 x 10^9/L) and (iii) severe (< 50 x 10^9/L). Individuals with a platelet count of 50 x 10^9/L or greater are asymptomatic (Gauer and Braun, 2012). Platelet count of 30 x 10^9/L or less can trigger bleeding from mild trauma, and a count < 10 x 10^9/L can cause spontaneous bleeding in mucosa, lungs, skin, central nervous system, gastrointestinal and genitourinary tract.

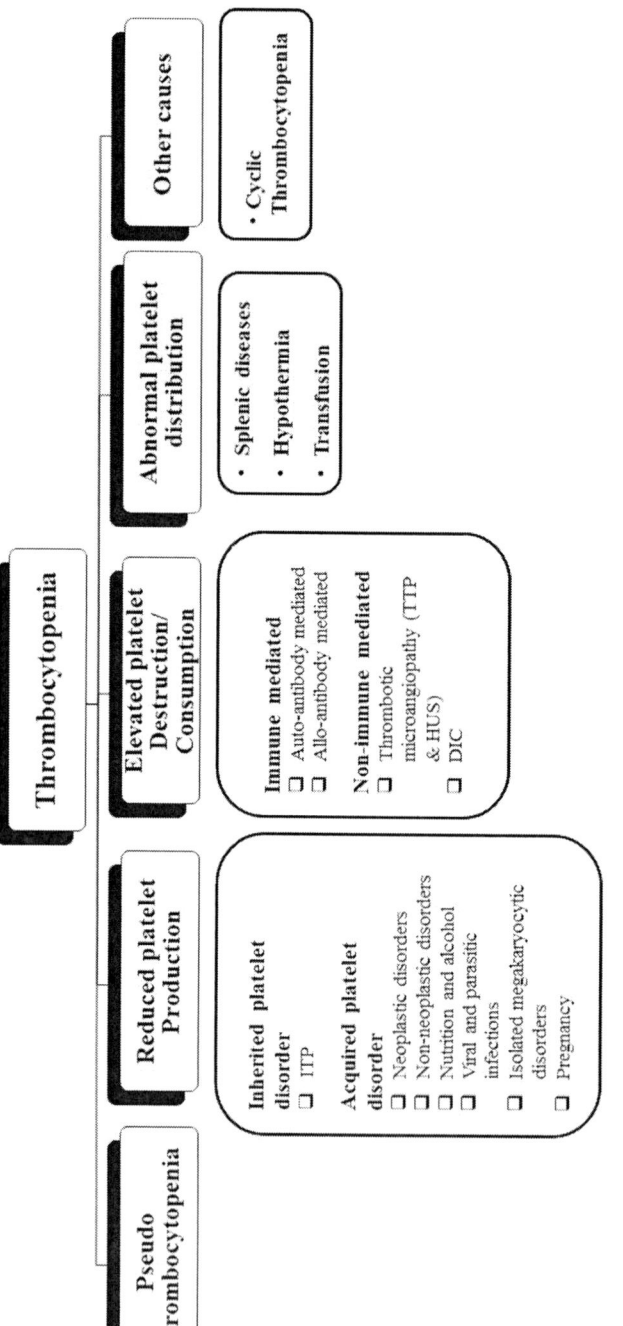

Figure 1. Classification of Thrombocytopenia.

Thrombocytopenia is caused due to one of the following three reasons:

I. Decreased platelet production in bone marrow
II. Increased platelet destruction/consumption
III. Elevated platelet sequestration in the spleen.

A cautious examination of peripheral blood film of the patient, along with the evaluation of marrow morphology is required to determine the etiology of thrombocytopenia.

Thrombocytopenia can be classified based on pathogenesis as follows (Figure 1):

1. Pseudo thrombocytopenia
2. Thrombocytopenia due to reduced platelet production.
   (a) Inherited platelet disorders
   (b) Acquired platelet disorders
3. Thrombocytopenia due to elevated platelet destruction
4. Thrombocytopenia due to abnormal platelet distribution
5. Thrombocytopenia due to other causes

PSEUDOTHROMBOCYTOPENIA (PTP)

Pseudothrombocytopenia is marked by incorrect low platelet count due to platelet aggregation in blood samples *in vitro*. These aggregates are formed when EDTA is used as an anticoagulant (Fang et al., 2015; Kamath et al., 2013; Shreiner and Bell, 1973). EDTA-induced PTP was first reported by Gowland et al., in 1969.

Incidence

EDTA-induced PTP has been widely reported since 1973. The prevalence of EDTA-induced PTP is around 0.1 - 2.0% in the hospitalized patients, 0.11 - 0.15% in out-patients and 0.013% in blood donors (Shreiner and Bell, 1973; Garcia Suarez et al., 1991; Maslanka et al., 2008; Froom and Barak, 2011).

Etiology

Aggregation of platelets in the presence of EDTA is due to the generation of autoantibodies against the glycoprotein (GP) IIb/IIIa on the platelet membrane. The chelating effect of EDTA on the Ca^{2+} ions exposes the epitope of GPIIb that is usually hidden in the GPIIb/IIIa complex. Binding of autoantibodies to the exposed epitope of GPIIb leads to platelet aggregation (Pegels et al., 1982; Casonato et al., 1994). Immunoglobulin (Ig) G, IgM as well as IgA; or the combination of IgG with IgM or IgA is involved in platelet aggregation (Payne and Pierre, 1984).

PTP is also observed in severely ill patients, associated with malignancy, cardiovascular, autoimmune, neoplastic, and atherosclerosis & liver-related conditions (Berkman et al., 1991; Isik et al., 2012).

Diagnosis and Prognosis

Misdiagnosis of PTP can lead to unnecessary diagnostics and treatments, which include bone marrow biopsy, splenectomy, steroid treatment and platelet transfusion (Payne and Pierre, 1984). Hence, it is important to consider PTP, when a low platelet count is detected. PTP can be confirmed by taking the patient's family history and bleeding tendency into consideration. Tubes containing anticoagulants (sodium citrate, heparin, and oxalate) other than EDTA must be used to re-examine the blood samples and a peripheral blood smear has to be prepared to identify PTP (Lombarts and de Kieviet, 1988).

Most of the antibodies involved in platelet aggregation react to cold conditions. Therefore, elevating the temperature of the blood sample by incubating at 37°C can aid to distinguish between PTP and true thrombocytopenia (Lombarts and de Kieviet, 1988; Bizzaro, 1995).

PTP can also be confirmed by adding kanamycin or amikacin to the blood samples collected using EDTA (Lombarts and de Kieviet, 1988). Sakurai et al., (1997) have reported that platelet aggregates dissociate when kanamycin is added within 30 min of blood collection with EDTA. However variations in morphology and complete blood count were insignificant (Figure 2).

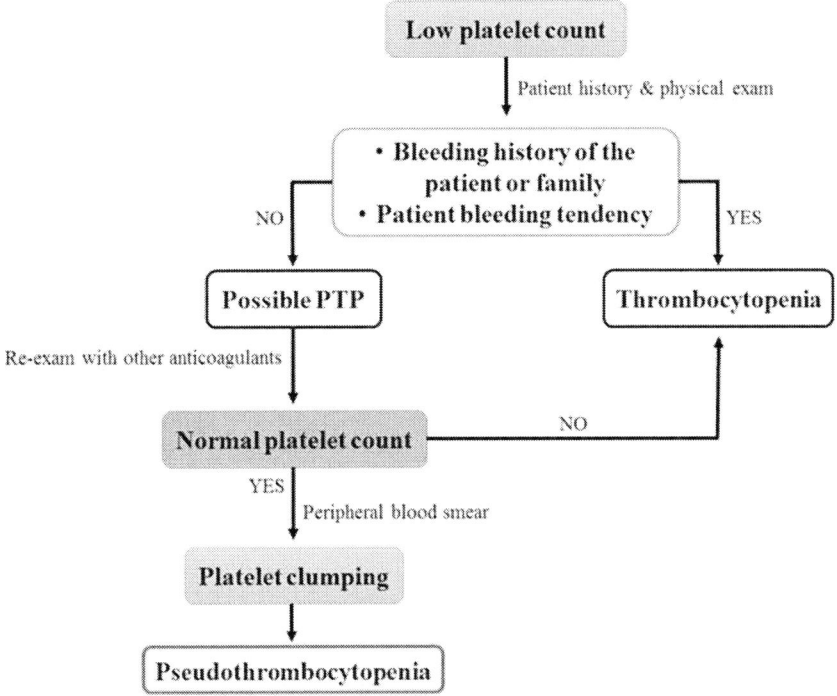

Figure 2. Identification of Pseudothrombocytopenia.

THROMBOCYTOPENIA DUE TO REDUCED PLATELET PRODUCTION

Platelet production can be reduced by the disorders that injure stem cells and/or prevent their proliferation. Multiple hematopoietic cell lines are generally affected, triggering anemia and leukopenia at varying degrees along with thrombocytopenia.

Thrombocytopenia due to reduced platelet production can be broadly classified into

(a) Inherited platelet disorders (IPDs)
(b) Acquired platelet disorders (APDs)

Inherited Platelet Disorders

Inherited platelet disorders are a heterogeneous group of disorders. For example, Bernard-Soulier syndrome, appear to be restricted to platelets (Diz-Kucukkaya and Lopez, 2013), whereas thrombocytopenia with absent radii (TAR) syndrome, forms a part of a complex pathology. Certain IPDs such as Glanzmann thrombasthenia, exhibit defective platelet function despite normal platelet count.

Etiology

A multitude of genes, ranging from transcription factors, cell surface receptors, cytokines, and signaling molecules to cell cycle regulators, molecular motors and cytoskeletal proteins, participate in differentiation of megakaryocytes and release of the platelets. Mutation in any one of these genes can reduce platelet production or their life span, or affect their morphology and function, leading to thrombocytopenia (Drachman, 2004).

Children may not demonstrate early bleeding tendency to detect severe forms of IPDs. Muco-cutaneous bleeding (purpura, epistaxis, and/or gingival bleeding) is a common symptom in IPD patients. Women may show signs of menorrhagia and bleeding during pregnancy and labor. Spontaneous life-threatening bleeding, such as intracranial hemorrhage, massive gastrointestinal or genitourinary bleeding, is uncommon (Gresele, 2015). A bleeding diathesis may only be diagnosed after an episode of excessive bleeding, such as surgery or trauma, in milder cases.

Diagnosis

Patients must be investigated for a possible inherited defect of platelet number or function under various circumstances. Diagnosis of IPD presents a significant challenge because of the heterogeneity of clinical and laboratory findings of the patients with the same disorder, even in the same family. It is vital to distinguish between patients with IPD and acquired platelet disorders [such as idiopathic thrombocytopenic purpura (ITP)], to avoid unnecessary or harmful treatments.

It might be helpful to obtain patient and family history since the majority of IPDs are inherited as autosomal recessive traits. The presence of skeletal, facial, audiologic, ocular, neurologic, cardiac, renal, and immune problems associated with platelet disorders may also suggest IPD (Balduini et al., 2013).

Patient History

Patient's history is an integral part of assessing the possible bleeding disorder and screen for a probable functional defect in platelets. Bleeding histories are subjective, variable and evolve throughout a person's lifetime. The bleeding manifestations of the reduced platelet number involve unexplained or extensive bruising; epistaxis; menorrhagia; oral cavity bleeding, and bleeding due to invasive procedures and dental extraction. A longstanding history of bleeding is highly significant in the diagnosis of IPD (Bolton-Maggs, 2006).

Family History

A family history compatible with the dominant forms of platelet disorders should be sought. Consanguineous partnerships increase the probability of a recessive platelet disorder. Investigation of family members might help to ascertain the diagnosis of inherited platelet disorder (Bolton-Maggs, 2006).

Examination

Laboratory evaluation of a potential IPD must include careful investigation of the blood film. This helps in assessing the nature of bleeding and exclude any underlying disease or diagnose a recognized syndrome. Traces of blood and protein have to be tested in urine samples (Bolton-Maggs, 2006).

Bleeding Time

Bleeding time does not correlate with the bleeding tendency in individual patients. A detailed bleeding history is considered to be a more valuable screening test (Burns & Lawrence, 1989; Peterson et al., 1998). Patients with collagen disorders usually have a normal bleeding time (De Paepe & Malfait, 2004). Prolonged bleeding time should prompt further investigation of platelet function.

Bleeding time can be used as a marker if other tests do not demonstrate any defects (Bolton-Maggs, 2006).

Platelet Function Analyser-100

Platelet function analyser-100 (PFA-100) measurements (Jilma, 2001; Michelson, 2007) will be significantly abnormal in Glanzmann thrombasthenia and Bernard–Soulier syndrome with closure times typically > 300 seconds on both ADP/collagen and adrenaline/collagen cartridges. Therefore, it can be used to screen patients to exclude these diagnoses. Patients with a history suggestive of a defective platelet function will need further investigation irrespective of PFA-100 analyses.

PFA-100 can be affected by platelet count, hematocrit, diet, aspirin, and von Willebrand factor (vWF) levels (Favaloro, 2001; Jilma, 2001). Therefore, a prolonged PFA-100 closure time needs to be investigated further with the measurements of vWF levels. Abnormal tests should be repeated to exclude a transient-acquired defect

Platelet Aggregation

Light transmission aggregometry (LTA) using different concentrations of ADP, collagen, ristocetin, epinephrine, and arachidonic acid is a gold standard in the diagnosis of IPDs. However, LTA may be normal in variant forms of IPDs, as well as in some patients with storage pool diseases. Quantifying the platelet nucleotide contents and release is recommended in the case of platelet granule deficiencies (Bolton-Maggs, 2006).

Flow Cytometry

Flow cytometric analysis is very informative in patients with platelet surface glycoprotein (GP) deficiencies such as Bernard-Soulier syndrome. Patients with pancytopenia or severe thrombocytopenia need marrow biopsy. Unfortunately, these tests may help diagnose only a small portion of IPD patients.

Flow cytometry is used to measure dense-granule content and release (Gordon et al., 1995; Wall et al., 1995). However, it is commonly used to assess GPIb/IX/V and GPIIb/IIIa in the diagnosis of Bernard–Soulier syndrome and Glanzmann thrombasthenia. Heterozygous states of these disorders can also be identified. GPIb/IX/V analysis in the heterozygous state is to be corrected for platelet size.

Flow cytometry requires small quantities of blood and therefore, is suitable to detect the defects in specific GPs, in the thrombocytopenic individuals (Bolton-Maggs, 2006).

Electron Microscopy

Electron microscopy is used to assess the defects in platelet granules and changes in the ultrastructure of platelets (e.g., MYH-9 defects) (White, 1969). The dense granule counts of the normal platelets can be used to determine the normal range and compare it with the patient sample [e.g., Dense granules are absent in Hermansky–Pudlak syndrome (HPS)].

Molecular Analysis

Molecular techniques such as western blotting, enzyme-linked immunosorbent assay (ELISA), or radioimmunoassay, are useful in both qualitative and quantitative analysis of specific platelet proteins. However, the results are inconclusive in 50% of IPD patients (Bolton-Maggs, 2006).

The genetic analysis of the patient can be fundamental in determining the molecular pathology; nevertheless, the traditional target gene method is complicated due to numerous genes. Several intriguing data regarding the genetic causes of IPD have been generated recently with the help of novel sequencing techniques for genetic investigations (Bunimov et al., 2013; Watson et al., 2013).

Prognosis and Management

IPD patients are informed about managing bleeding episodes. Preventive measures include informing the patients and their family regarding the maintenance of dental hygiene, avoiding activities that can lead to injuries, as well as refraining from the consumption of antiplatelet drugs. Compression and use of tranexaminic acid-embedded gauze, fibrin sealants or gelatin sponges can be used to control the local bleeding (such as epistaxis and gingival bleeding). Educating female patients about menorrhagia, and the use of oral contraceptives, antifibrinolytic drugs, and intrauterine hormonal devices can help to reduce the uterine bleeding. Bleeding in patients with BSS and GT can be decreased using desmopressin, which helps to elevate the vWF concentration in plasma. Iron supplementation is the treatment for patients with chronic bleeding issues. Platelet and erythrocyte transfusions are a regular practice for patients undergoing surgery or severe bleeding. However, such patients might develop antibodies against platelets. Recombinant factor VIIa is used in patients with life-threatening bleeding, or with alloantibodies against platelets (Seligsohn, 2012; Diz-Kucukkaya and Lopez; 2013).

Recently thrombopoietin (TPO) mimetics such as romiplostim and eltrombopag, are being used to elevate the platelet count in patients with chronic and refractory ITP.

TPO mimetics can also be used to increase platelet count in patients before surgery (Balduini et al., 2011).

Hematopoietic stem cell (HSC) transplantation and gene therapy are the recommended treatments for patients with severe IPD. However, the risk/benefit ratio has to be evaluated carefully in each patient. Transplantation is advisable to patients with an HLA-identical sibling donor. Matched-unrelated donor transplant is suggested for patients with excellent HLA match, using T-cell depletion (Filipovich et al., 2001; Alamelu and Liesner, 2010).

Acquired Platelet Disorders

An acquired disorder of platelet numbers can be due to marrow diseases, which can result in pancytopenia or isolated megakaryocytic disorder. Other causes include neoplastic and non-neoplastic diseases, nutrition and alcohol, viral infections and parasites, and isolated megakaryocytic disorders.

Neoplastic Disorders

Acute disorders such as myeloid and lymphoid leukemia, and chronic lymphoproliferative disorders (such as hairy-cell leukemia, T-cell leukemia and lymphoma, chronic lymphoid leukemia and non-Hodgkin's lymphoma) are the primary marrow neoplastic disorders associated with thrombocytopenia (Dubansky et al., 1989; Kim et al., 2014). The presence of pancytopenia and malignant cells in circulation prompts the diagnosis. Metastatic malignancies and infectious organisms can infiltrate the bone marrow, leading to thrombocytopenia.

Non-Neoplastic Disorders

Non-neoplastic hypo proliferative stem cell disorders (for instance, non-immune drug-induced aplasia, chemotherapy-induced pancytopenia, and aplastic anemia) are associated with thrombocytopenia (Nahirniak et al., 2015).

Patients with myelodysplasia (MDS) frequently present with acquired thrombocytopenia (Hellstrom-Lindberg et al., 1999).

Nutrition and Alcohol

Deficiency of vitamin B12 (folic acid) is known to cause a reduced production of platelets. Vitamin B12 deficiency leads to defective synthesis of DNA in megakaryocytes, which results in lower ploidy and reduced number of giant platelets (Bessman, 1984). The large platelets, combined with the large, heterogeneous RBCs generally indicate a vitamin B12 deficiency. Iron replacement therapy also causes a transient decrease in platelet count, since the stem cells shift to erythropoiesis.

Gastric bypass surgery patients have copper deficiency leading to anemia, leukopenia, and thrombocytopenia and also resemble vitamin B12 deficiency, when associated with neurologic deficits. Elevated ring sideroblasts and dysplastic precursor cells in the marrow smears lead to misdiagnosis as myelodysplastic syndrome (Agnotti et al., 2008; Green, 2012).

Severe and prolonged alcohol consumption is also known to cause acquired thrombocytopenia (Ballard, 2015), particularly a decrease in megakaryocytes, along with normoblast and promyelocyte vacuolation (Lindenbaum and Lieber, 1969). Chronic alcoholism may cause thrombocytopenia by alcoholic liver cirrhosis (splenomegaly and TPO deficiency), alcohol-induced marrow suppression and the folic acid deficiency (Sullivan et al., 1977; Michot and Gut, 1987; Smith et al., 1992; Wolber and Jelkmann, 2002; Latvala et al., 2004). Discontinuation of alcohol consumption elevates the platelet count after 2-3 days.

Viral and Parasitic Infections

Some viral infections (such as the human immunodeficiency virus (HIV), Epstein-Barr, hepatitis B, hepatitis C, parvovirus B19, cytomegalovirus, Mumps, rubella and varicella-zoster) reduce platelet number, and can cause acute thrombocytopenia by suppression of bone marrow (Flaujac et al., 2010; Gauer and Braun, 2012).

Thrombocytopenia is more prevalent in untreated HIV infections due to the decreased life span of platelets, HIV-infected megakaryocytes, and increased splenic sequestration (Cole et al., 1998; Kaushansky et al., 2016). Declined TPO production from the liver, autoimmune component, and inhibition of bone marrow heightens the risk of thrombocytopenia in patients with hepatitis C infection (Olariu et al., 2010). Furthermore, patients recover from thrombocytopenia in viral infections like H1N1 (Gauer and Braun, 2012).

Parasitic infections, such as malaria, can affect the platelet numbers by increased sequestration, decreased platelet life span, and immune-mediated destruction (Nithish et al., 2011). Furthermore, the platelets also bind to the infected RBCs, thereby leading to high platelet consumption (McMorran et al., 2012).

Isolated Megakaryocytic Disorders

Isolated thrombocytopenia, accompanied by complete megakaryocyte reduction, is an indication of acquired megakaryocytic thrombocytopenia. Sometimes this can progress to MDS or aplastic anemia. The pathogenesis involves an intrinsic defect in stem cells, autoantibodies against the megakaryocyte precursors and T-cell mediated megakaryocyte development suppression. However, the immunosuppressive treatments have shown to aid the patients, implying the role of immune system in pathophysiology (Bakchoul and Greinacher, 2016).

THROMBOCYTOPENIA DUE TO ELEVATED PLATELET DESTRUCTION

The causes of platelet destruction can be immune-mediated or non-immune mediated. Figure 3 depicts the further classification of thrombocytopenia due to elevated platelet destruction.

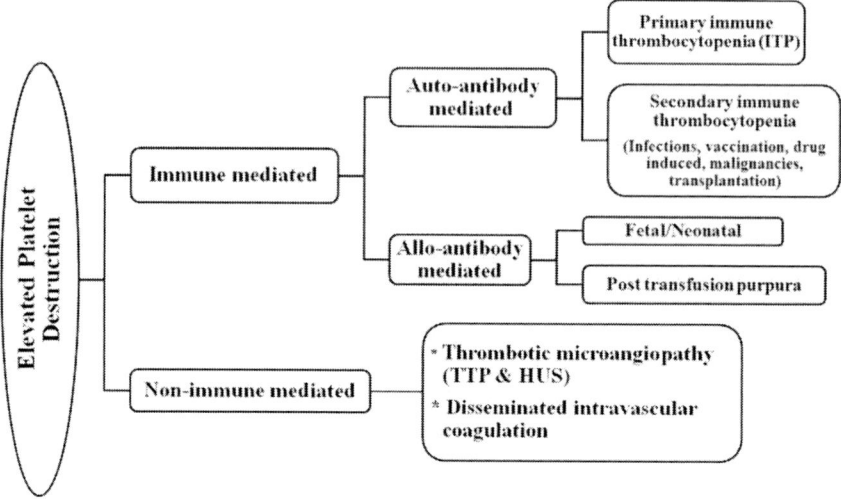

Figure 3. Thrombocytopenia due to elevated platelet destruction.

Immune-Mediated Thrombocytopenia

Immune thrombocytopenia can be auto-antibody mediated or allo-antibody mediated. Auto-antibody mediated thrombocytopenia includes primary thrombocytopenia and secondary thrombocytopenia.

Auto-Antibody Mediated Thrombocytopenia

Primary Immune Thrombocytopenia

Etiology

Idiopathic thrombocytopenic purpura (ITP) (also called autoimmune thrombocytopenia purpura) is a common cause of isolated thrombocytopenia. The characteristic feature of ITP includes immune-mediated platelet destruction and damaged platelet production. ITP can also develop after viral infection or vaccination in children and is generally acute.

Table 1. Therapeutics in ITP (Kaushansky et al., 2016)

Treatment	Remarks
Glucocorticoids	• Increase platelet count by inhibiting phagocytosis of Ab-coated platelets. • A decrease in auto-Ab production. • Improve platelet production in the marrow. • Side effects: Facial Swelling, weight gain, folliculitis, hypertension, hyperglycemia, cataracts, opportunistic infections. • Include prednisone, dexamethasone & methylprednisolone.
Splenectomy	• Effective treatment • Second-line therapy. • Spleen: Major site for anti-platelet Ab synthesis & Ab-coated platelet destruction. • Decrease Ab production & platelet destruction. • Effective with patients with Ab-medicated platelet destruction. • The highest cure rate in ITP patients. • Invasive & permanent loss of an organ. • Increased risk of bacterial infection, bleeding & thrombosis.
Intravenous Immunoglobulin (IVIg)	• Effective for IPT. • Rapidly elevates Platelet counts. • Blocks Fc receptors of macrophages & slows the clearance of Ab-coated platelets, cytokine modulation, immune modulation & complement neutralization. • Expensive. • Adverse Effects: headache, backache, nausea, fever, aseptic meningitis, hepatitis, pulmonary insufficiency, renal failure & thrombosis. • Reserved for patients with life-threatening bleeding or first-line therapy fails.
Anti-(Rh)D	• Polyclonal γ-globulin with high Ab titers against Rh(D) Ag of RBCs. • Administered intravenously. • Binds to Rh-positive RBCs which are then destroyed in the spleen. • More Ab-coated platelets survive since the splenic Fc receptors are blocked. • Not effective in patients who have undergone splenectomy in Rh-negative patients. • Not recommended in patients with a positive direct antiglobulin test. • Given as a single dose of 50 - 100mcg/kg by IV infusion over 3-5mins. • Adverse effects: headache, chills, fevers, asthenia, diarrhea, abdominal pain, dizziness, vomiting & myalgia. • Immediate anaphylactic reactions & both Type-I & Type III hypersensitivity reactions noted in a few patients. • Increase platelet count within a week.
Rituximab	• Target the B cells. • A chimeric monoclonal Ab against CD20, which binds B cells & cause Fc-mediated lysis, thereby depleting these cells from blood, lymph nodes & marrow. • Rapidly depletes B-cells in patients with the autoimmune disease. • Severe infusion-related reactions, increased risk of infections.

Table 1. (Continued)

Treatment	Remarks
Thrombopoietin (TPO) receptor agonists	• Stimulate platelet production by increased megakaryopoiesis. • Romiplostim & eltrombopag stimulate platelet production.
Romiplostim	• Polypeptide carrying 4 copies of a 14-amino acid TPO receptors binding peptide fused to an immunoglobin scaffold. • Binds to the TPO-binding site of the TPO receptor with high affinity. • Induces megakaryocyte proliferation & differentiation by activating Janus-type tyrosine Kinase (JAK)-signal transducer & activator of transcription (STAT) & mitogen-activated protein (MAP) kinase pathways.
Eltrombopag	• A 422 Da non-peptide molecule. • Binds to the transmembrane domain of the TPO receptors & triggers megakaryocytes growth & differentiation, increase platelet production. • Does not compete with TPO binding.
New TPO receptor agonists	• Avotrombopag: oral non-peptide TPO agonist. • Binds to the transmembrane domain of the TPO receptors & increase platelet count. • Side effects: mild headache, narcopharyngitis, arthralgia, fatigue, nausea & diarrhea. • Can induce extreme thrombocytosis.
Azathioprine	• A purine analog that is converted to 6-mercaptopurine following gastrointestinal absorption. • Works by suppressing the immune response. • Can be used in pregnancy if necessary. • Adverse Effects: Marrow suppression & an increased risk of secondary malignancy.
Cyclophosphamide	• Alkylating drug used orally or parenterally in patients with refectory ITP. • Beneficial action linked to immunosuppression. • Complications: marrow suppression, infertility, hemorrhagic cystitis, alopecia & secondary malignancy.
Cyclosporine	• Immunosuppression drug inhibiting T-cell function. • Used to prevent rejections in patients with organ transplantation. • Side Effects: fever, opportunistic infections, diarrhea, gingival hyperplasia, peptic ulcer, renal dysfunction, pancreatitis, hypertension, elevated liver enzymes, convulsion, peripheral neuropathy, secondary malignancy & hirsutism.
Danazol	• Synthetic androgen to treat patients with refectory ITP. • Decrease Fc receptors numbers on phagocytic cells by antagonizing the effects of estrogen. • Should not be given to pregnant women. • Side effects: weight gain, seborrhea, fluid retention, secondary amenorrhea, hirsutism acne, vocal changes, headache, hepatic toxicity, lethargy, myalgia & cholesterol spectrum abnormalities.

Treatment	Remarks
Dapsone	• Increased platelet count in patients with chronic, refectory or persistent ITP. • Discontinuation of dapsone results in platelet count to baseline. • Side Effects: nausea, skin rashes, headache, cholestasis, hepatitis, methemoglobinemia & dose-dependent hemolysis. • Should not be given to the patients with G6PD deficiency.
Vinca alkaloids	• Vincristine & vinblastine transiently increase platelet count.
	• Administered by Bolus injection. • Bind to platelet microtubules & are transported to spleen, where they inhibit phagocytic functions of splenic macrophages. • May also stimulate megakaryopoiesis. • Complications: peripheral neuropathy, jaw pain, neutropenia, constipation & alopecia.
Other therapies	• Eradication therapy for ITP patients with *Helicobacter pylori* infection. • Interferon-α, ascorbic acid, immunoadsorption with staphylococcal protein A, plasmapheresis & colchicine for patients with refectory ITP.

However, the condition is chronic in adults and does not resolve spontaneously (Lusher and Iyer, 1977).

The most common antiplatelet antibodies are IgG, along with IgM and IgA. Spleen, liver and marrow destroy the antibody-coated platelets after they bind to the tissue macrophages via Fc_γ receptor (Shulman et al., 1965; McMillan, 2000; Cines and Blanchette, 2002). Patients with ITP have been reported to have various abnormalities in cell-mediated immunity [abnormalities in antigen (Ag) presenting cells, cytokine release, and T-lymphocytes].

T-cells differentiate into T-helper 1 (Th1), T-helper 2 (Th2), Th17 and T-regulatory (T_{reg}) cells. T_{reg} cells inhibit autoimmune responses and thereby play a vital role in self-tolerance. The Th1 and Th17 cells are found to be up-regulated in ITP patients, whereas there is a decrease in the T_{reg} cell numbers and suppressor functions (Stasi et al., 2001; Liu et al., 2007; McKenzie et al., 2013). This imbalance induces an autoimmune response against platelets. The CD8+ cytotoxic T-cells are also reported to cause cell-mediated destruction of platelets and megakaryocytes, thereby adding to the pathogenesis of ITP (Zhang et al., 2006; Li et al., 2007).

Diagnosis

A specific laboratory test to identify ITP is not available. ITP diagnosis requires the elimination of the other causes of thrombocytopenia.

Prognosis

Treatment for ITP includes both pharmacological and surgical approaches. Table 1 summarizes the various treatment methods available for ITP.

Secondary Immune Thrombocytopenia

Secondary immune thrombocytopenia is characterized by immune-mediated platelet destruction due to conditions such as systemic lupus erythematosus (SLE), anti-phospholipid syndrome (APS), and some infections. Secondary ITP can be a symptom of an existing illness or develop concurrently with the disease or therapies. Secondary ITP is not predominantly severe, however lower platelet counts can pose the risk of bleeding. Treatment strategy must be designed specifically to an individual patient.

Anti-Phospholipid Syndrome (APS)

The characteristic features of APS include recurrent arterial and venous thrombosis, or both; and morbidity in the presence of anti-phospholipid antibodies (APLAs) during pregnancy (Miyakis et al., 2006).

Etiology

Although the pathogenesis of ITP during APS is unclear, the possible mechanisms involve platelet destruction by APLAs, antibodies against the glycoproteins, by complement, and platelet aggregation and consumption. APLAs are reported to bind to platelet membranes and destroy platelets. However, patients with APS rarely present with bleeding complications (Galli et al., 1994).

Prognosis

Anti-glycoprotein antibodies reduce, and the platelet numbers elevate; whereas, the APLAs do not decrease with immunosuppressive treatments in APS patients. Only 15% of patients respond to glucocorticoids (Galli et al., 1996). Patients presented with severe bleeding are treated with intravenous immunoglobulin (IVIg), immunosuppressive drugs (cyclophosphamide and azathioprine), or splenectomy (Stasi et al., 1995; Font et al., 2000).

Systemic Lupus Erythematosus (SLE)

Women of childbearing age are primarily afflicted by SLE, which is a complex autoimmune disease. Any tissue can be affected by the autoimmune attack and is not explicit to any organ (Tan et al., 1982; Hochberg, 1997).

Etiology

SLE patients commonly exhibit thrombocytopenia. Platelet destruction [disseminated intravascular coagulation (DIC), thrombotic thrombocytopenic purpura (TTC), sepsis, hemolytic uremic syndrome, drugs], hypersplenism, ineffective hematopoiesis, and marrow hypoplasia can lead to thrombocytopenia in patients with SLE. Severe thrombocytopenia is uncommon; however, fatal cerebral, pulmonary, and gastrointestinal bleeding have been reported. Antiplatelet antibodies against integrins is the leading cause of platelet destruction in SLE patients (Rabinowitz and Dameshek, 1960; Pujol et al., 1995; Michel et al., 2002).

Prognosis

Thrombocytopenia in SLE patients ranges from mild and manageable to severe and fatal. Glucocorticoids are the general first-line treatment (Boumpas et al., 1990; Arnal et al., 2002). Patients with severe bleeding are treated with IVIg (Maier et al., 1990; Cohen and Li, 1991). Rituximab (targeting B-cells) is recommended for refractory SLE patients with severe thrombocytopenia and nephritis (Ding et al., 2008).

Splenectomy is not preferred in SLE patients since it can augment the thrombotic complications and risk of infection (Zhou et al., 2013).

Infectious Diseases

Patients suffering from viral, fungal, bacterial, or parasitic infections also present with thrombocytopenia.

Etiology and Prognosis

Elevated immune destruction of platelets, decreased platelet production in the marrow, and microangiopathy may lead to thrombocytopenia during infections. The drugs given to treat the underlying infections also cause thrombocytopenia. Viral infections mainly give rise to secondary ITP. Thrombocytopenia can exist with other symptoms in virus-infected patients of rubella, infectious mononucleosis, and mumps. However, thrombocytopenia resolves in 2-8 weeks in such cases.

Thrombocytopenia during HIV infection can be due to various reasons such as, immune complex-related platelet destruction, reduced platelet production, sequestration by spleen and platelet consumption in combination with thrombotic thrombocytopenic purpura (TTP) (Dominguez et al., 1994; Louache and Vainchenker, 1994).

Thrombocytopenia in adults can also be caused by hepatitis C virus (HCV). The mechanisms involve immune-mediated platelet destruction, declined TPO levels with liver insufficiency and drugs. Immune-mediated platelet destruction includes binding of free, as well as IgG-complexed HCV to the platelets (Stasi, 2001). Treatment comprises of antiviral therapy [pegylated interferon (INF) and ribavirin] to diminish the viral load. Nevertheless, the platelet count can remain unaffected or can further decrease post-treatment. Severe thrombocytopenia elevates the risk of bleeding and interferes with HCV treatment. IVIg is advised as first-line therapy during such conditions (Neunert et al., 2011). HCV patients with liver cirrhosis possess the risk of abdominal thrombosis when treated with TPO receptor agonists (Cuker, 2010).

Table 2. List of drugs reported to induce thrombocytopenia (Kenney and Stack, 2009; Erkurt et al., 2012)

Category	Drugs
Antiarrhythmic	• Procainamide
Antidepressants	• Amitriptyline • Desipramine • Mianserine
Anti-epileptic/Anticonvolusant	• Carbamazepine • Phenytoin • Valproate
Anti-GPIIb/IIIa	• Abciximab • Eptifibatide • Lotrafiban • Tirofiban
Anti-histamines	• Antazoline • Chlorpheniramine
Antihypertensive	• Hydrochlorothiazide Antihypertensive
Anti-inflammatory	• Acetaminofphen • Diclofenac • Fenoprofen and iboprofen • Indomethacin • Meclofenamate • Mefanimic acid • Naproxen • Oxyphenylbutazone • Phenylbutazone • Piroxicam • Salicylates (Aspirin, Diflunisal, Aminosalicylate, Sulfosalazine) • Sulfasalazine • Sulindac
Antimicrobial	• Cephalosporins • Para-aminosalicylic acid • Trimethoprim-sulfamethoxazole
Antimony containing drugs	• Sodium stibogluconate • Stibophen
Antirheumatic	• Gold salts
Antiviral	• Acyclovir • Interferon
Benzodiazepines	• Diazepam
Cardiac medications and diuretics	• Acetazolamide • Alpha-methyldopa • Alprenolol • Amiodarone • Captopril • Chlorthiazide • Diazoxide • Digitoxin

Table 2. (Continued)

Category	Drugs
Cardiac medications and diuretics	• Digoxin • Furosemide • Hydrochlorthiazide • Oxprenolol • Procainamide • Spironolactone
Corticosteroid	• Prednisone
Growth factor receptor blocker	• Suramin
H2 antagonists	• Cimetidine • Ranitidine
Heparin	• Low molecular weight • Unfractionated (Regular)
Miscellanous drugs	• Actinomycin-D • Aminoglutethimide • Chlorpropamide • Danazole • Desferrioxamine • Etretinate • Glibenclamide • Iodinated contrast agents • Levamizole • Lidocaine • Measales-mumps-rubella vaccine • Morphine • Papaverine • Quinine and quinidine • Retinoids, isotretinoin • Sulfonylurea drugs • Tamoxifen • Ticlopidine

Drug-Induced Thrombocytopenia (DIT)

Vipan, in 1865, first described the occurrence of thrombocytopenia due to quinine. Since then, numerous drugs have been reported to cause thrombocytopenia. DIT is not severe and affects only a few patients; however, it can sometimes be fatal. Both genetic, as well as environmental factors, influence the susceptibility of patients to drugs.

Heparin-induced thrombocytopenia (HIT) is the most severe form of DIT. The antibodies recognize the exposed neopeptide on platelet factor 4 (PF4) when it binds to heparin. As a result, the platelets are activated, initiating the coagulation cascade, which leads to venous and arterial thrombosis.

Etiology

The drug-dependent antibodies (DDAbs) cause immune destruction of platelets, resulting in thrombocytopenia (Aster and Bougie, 2007). The presence of the causative drug in the system triggers different immune mechanisms, and in turn, the DDAbs bind to platelets via platelet Fc$_\gamma$ receptor (in HIT) or F$_{ab}$ region (other drugs) (Christie et al., 1985). The platelet surface GPs (GPIb/IX/V or integrin $\alpha_{IIb}\beta_3$) are the main target antigens (Visentin et al., 1991; Curtis et al., 1994; Lopez et al., 1995). DIT is more frequently caused by sulfonamides, quinine, and quinidine. A list of various drugs that can induce thrombocytopenia is given in Table 2, and the different mechanisms of DIT are listed in Table 3.

Diagnosis

There are five criteria for DIT diagnosis (Arnold et al., 2013):

(i) Exposure to the candidate drug preceded thrombocytopenia
(ii) Discontinuation of the drug led to complete and sustained recovery
(iii) Only the candidate drug was used before the occurrence of thrombocytopenia
(iv) Exclusion of other causes for thrombocytopenia
(v) Recurrent thrombocytopenia was observed upon re-exposure to the candidate drug.

DIT usually occurs in patients who are often hospitalized and take multiple medications. There are laboratory tests that can detect the antibodies that bind to platelets in the presence of drugs or their metabolites. Other techniques like flow cytometry (Curtis et al., 1994), solid phase red cell adherence assays (Leach et al., 1997) and monoclonal antibody-specific immobilization of platelet antigen (MAIPA) (Nieminen and Kekomaki, 1992) can also be used to detect DDAbs, however their results are not clinically validated.

Table 3. Mechanisms of Drug-induced Thrombocytopenia. (Kaushansky et al., 2016)

Type	Mechanisms
Quinin type	• Drug-dependent antibodies attach to platelets, only in the presence of sensitizing drug. • The antibodies bind to platelet GP epitopes.
Neoepitope	• Conformational changes are activated in protein structure due to the interaction between drug and membrane proteins this elicits DDAbs.
Hapten	• Small molecules covalently couple with large carrier proteins (GPIIb-IIIa) and the DDAbs are induced.
Drug-specific	• The F_{ab} fragment of the chimeric monoclonal antibody specific for GPIIIa is recognized by the drug-specific antibodies.
Autoantibody	• Drug exposure induces the antibodies, but antibody binding to the platelets does not depend on the presence of the drug.

Prognosis

Withdrawal of the offending drug is the most important therapeutic measure. Patients with severe thrombocytopenia and bleeding are treated with IVIg. Platelet transfusions and a high dose of parenteral methylprednisolone are the recommended treatments in situations of emergency (George and Saucerman, 1988).

Allo-Antibody Mediated Thrombocytopenia

Fetal/Neonatal Thrombocytopenia

After the first trimester, the fetal platelet count becomes normal (> 150 x 10^9/L), which continues throughout gestation. Nevertheless, pre-term infants possess the risk of thrombocytopenia. Fetal/Neonatal allo-immune thrombocytopenia (NAIT) is a life-threatening condition in neonates if it is severe (< 50 x 10^9/L). Therefore once detected, it needs careful handling due to increased risk of bleeding (Chakravorty and Roberts, 2011).

Etiology

The maternal allo-antibodies (IgG) transfer through placenta and bind to the paternal fetal platelet antigens and trigger their clearance from circulation leading to thrombocytopenia. NAIT is similar to the hemolytic anemia in the neonates. The maternal antibodies destroy the fetal cells (antigen positive), causing fetal/neonatal morbidity and mortality. At least, 40 - 60% of the firstborns are affected by NAIT (Letsky and Greaves, 1996). Antiplatelet antibodies can also transfer to the babies born from mothers suffering from ITP. Maternal ITP does not lead to severe bleeding in the fetus, although NAIT can lead to intracranial hemorrhage (Kelton et al., 1982).

Diagnosis

The maternal platelet count is found to be normal in NAIT, unlike maternal ITP, and serves as a critical diagnostic tool. Confirmatory test for NAIT involves detection of maternal antibody, circulating in the fetus, against the paternal platelet antigen. Assays such as flow cytometry and ELISA can be used to detect the maternal antibody. Weak or mixtures of antibodies can also be detected by employing MAIPA along with the above methods (Kiefel et al., 1987; Kaplan, 1998).

Prognosis

The treatment cannot be delayed in case of bleeding and severe thrombocytopenia at birth, due to the risk of hemorrhage in the infant. Immediate transfusion of platelets that will not be cleared by the maternal allo-antibodies is recommended. Therefore, the mother is considered as the best donor (Murphy et al., 1999). Nevertheless, the maternal allo-antibodies will have to be removed by washing before transfusion. When the compatible platelets are unavailable, treatment with random-donor platelet concentrates are recommended (Win, 1996), although low median platelet count elevations have been reported (Spencer and Burrows, 2001).

Transfusion of random platelets in combination with perfused IVIgG is recommended, if there are no other alternatives. However, it must be noted that platelet transfusion cannot be substituted with IVIgG since it can lead to significant bleeding due to the delayed response of 12-18 h (Massey et al., 1987).

Post Transfusion Purpura (PTP)

Severe thrombocytopenia is observed in patients, 5-10 days after transfusion (Warkentin and Smith, 1997). PTP is a rare condition, though it is severe. A sensitized person develops a potent human platelet-specific alloantibody (HPA) (usually anti-HPA-1a). HPA-1a negative patient receiving a transfusion from HPA-1a positive donor leads to the production of anti-HPA-1a (Minchinton et al., 1990; Lucas et al., 1997), and destruction of both donor and patient platelets, resulting in severe thrombocytopenia. Nevertheless, the condition returns to normal in 3-4 weeks. Treatments include plasmapheresis or IVIg under severe conditions (Santoso and Kiefel, 2001).

Non-Immune-Mediated Thrombocytopenia

Thrombotic Microangiopathy

Microangiopathic hemolytic anemia (MAHA), thrombocytopenia and microvascular thrombi are the characteristic features of thrombotic microangiopathy. It is classified into TTP and hemolytic uremic syndrome (HUS) (George, 2000; Ruggenenti et al., 2001). Patients with TTP have a deficiency of the vWF cleaving protease ADAMTS13 (a disintegrin-like and metalloprotease with thrombospondin type 1 motif); however, it is normal in HUS patients (Furlan et al., 1998; Bianchi et al., 2002).

Thrombotic Thrombocytopenic Purpura (TTP)

TTP is more common in females (Ridolfi and Bell, 1981). African ancestry and obesity are the other risk factors (Nicol et al., 2003; Vesely et al., 2003). Platelet count of $< 20 \times 10^9/L$ can cause severe thrombocytopenia, along with muco-cutaneous bleeding (Ridolfi and Bell, 1981). Fibrinogen levels, partial thromboplastin time, and prothrombin time (PT) are normal. However, few patients demonstrate an increase in fibrin/fibrinogen degradation products. The coagulation and fibrinolytic pathways are also activated.

TTP can be inherited or acquired. Inherited TTP is an autosomal recessive disorder, where patients have recurring episodes of thrombotic microangiopathy, associated with vaccination, infections or surgeries, in early childhood (Furlan and Lammle, 2001; Veyradier et al., 2003). Acquired TTP is associated with ADAMTS13 deficiency due to inhibitory antibodies IgG (Furlan et al., 1998; Tsai and Lian, 1998). Stress is shown to play a significant role in aggravating both inherited and acquired TTP. Alterations in pro- and anti-fibrinolytic factors, enormous procoagulant activity by circulating endothelial microparticles, mutations or deficiencies of natural anticoagulant mechanisms or the effects of vasoregulatory substances (endothelin and prostacyclin) can also contribute to the pathogenesis of TTP.

The mortality rate of TTP is 90% without therapy. Plasma exchange is the recommended treatment for TTP. Initially, 1.5 volumes of plasma are exchanged per day. The volume and frequency of the exchange are varied based on the patient's response and extended until the symptoms resolve (Scully et al., 2012). Rituximab is suggested to the patients who do not respond to plasma exchange (Scully, 2012). Other treatments include steroids (methylprednisolone), immunomodulators (vincristine and cyclosporine A), N-acetylcysteine, and bortezomib (Joly et al., 2017). Splenectomy is recommended only to the refractory patients (Essein et al., 2003).

Hemolytic Uremic Syndrome (HUS)

HUS has different variants. HUS in children is due to infection with *E. coli* that produces verotoxin. Bloody diarrhea is a significant symptom and hence, is referred to as D(C) or childhood HUS. MAHA, thrombocytopenia, and renal failure are the features of sporadic HUS, and is common in adults. It is not associated with bloody diarrhea and is referred to as D(K) or adult HUS (Berns et al., 1992). Genetic deficiencies of complement regulatory proteins lead to inherited HUS, which is a familial disorder (Neild, 1987).

IgG depletion using protein G columns improve the severe neurological symptoms and renal dysfunction in HUS (Greinacher et al., 2011). Adult

HUS patients are treated with eculizumab, to block complement C5 (Lapeyraque et al., 2011)

Disseminated Intravascular Coagulation (DIC)

DIC can arise due to organ destruction, severe hepatic failure, sepsis, tumors, trauma, and other causes, leading to intravascular activation of coagulation (Gando et al., 2013). Activation of coagulation and hyperfibrinolysis are the results of a massive inflammatory response, along with tissue factor release in the patient. Thrombocytopenia observed in DIC is the result of elevated platelet consumption in the process, and patients commonly display bleeding and microthrombi.

DIC must be diagnosed if the causative factor exists, and is validated by laboratory results. Tests such as PT and activated partial thromboplastin time (aPTT) are used to screen the extent of coagulation factor consumption and activation. Fibrin formation can be indirectly measured by assessing the levels of D-dimers in blood. Thrombocytopenia, prolonged PT, aPTT, elevated fibrin degradation products (FDPs), and low fibrinogen are the abnormalities found in DIC (Wilde et al., 1989).

Managing the primary cause of DIC is often suggested as treatment (Richey et al., 1995). Thrombosis prophylaxis is considered standard treatment, and such patients (with mild DIC and no bleeding evidence) do not require additional treatments (Saito, 1993; Kobayashi et al., 2001). According to experts, platelets are transfused to the patients with active hemorrhage, or if the count is below $50 \times 10^9/L$.

THROMBOCYTOPENIA DUE TO ABNORMAL PLATELET DISTRIBUTION

Splenomegaly and Hypersplenism

One-third of total platelets are generally pooled in the spleen at any given time under normal conditions. Splenomegaly is when approximately

90% of the total platelets are pooled in the spleen, causing thrombocytopenia (Aster, 1972; Wadenvik et al., 1987). However, platelet production is normal in splenomegaly patients (Aster, 1972).

Disorders such as congestive splenomegaly, cirrhosis, and chronic liver disease with portal hypertension are the most common causes of thrombocytopenia in splenomegaly. Thrombocytopenia can be a result of reduced production of TPO in the liver, apart from splenic pooling.

The symptoms of splenic pooling are related to that of the primary disorder. Coagulation abnormality due to the underlying liver disorder can cause bleeding. The platelet count returns to normal upon splenectomy or correction of portal hypertension by surgery (Aster, 1972; Lawrence et al., 1995). Platelet transfusions are not recommended as the transfused platelets are also sequestered in the spleen and do not significantly elevate the platelet count.

Hypersplenism is a condition where pooling follows increased destruction of platelets, erythrocytes, and leucocytes. Thrombocytopenia aggravates in patients with severe cirrhosis due to impaired production of TPO and cirrhosis associated DIC. Treatment of the underlying disease is the recommended approach (Aster, 1966; Jandl et al., 1967; McCormick and Murphy, 2000).

Hypothermia

Platelet activation and thrombocytopenia are the side effects of hypothermia, i.e., body temperature below 25°C (Villalobos et al., 1958). Hypothermia leads to an elevation in platelet sequestration and is used to prevent ischemia-related organ damage (Easterbrook and Davis, 1985; Vella et al., 1988).

GPIb complex clustering and rearrangement of its carbohydrate chains are triggered under hypothermic conditions, and act as a ligand for integrin $\alpha_M\beta_2$ on the macrophages, leading to platelet clearance by hepatic macrophages (Hoffmeister et al., 2003). ADP is also known to play a major role in hypothermia-induced platelet activation. However, the condition is only transient and is reversed upon restoring the body temperature (Pina-Cabral et al., 1985); hence, no treatment is required.

Massive Transfusion

Massive transfusion is the transfusion of one volume of whole blood or more than 10 units of packed RBCs in 24 hours and transfusion of more than 4 units of packed RBCs over 1 hour (Sihler and Napolitano, 2009). Patients with uncontrolled or heavy bleeding are treated with massive transfusion, which can lead to mild or severe thrombocytopenia in some patients. Platelet dilution due to transfusion of packed RBCs, or DIC (blood loss due to disease or after trauma) are the causes of transfusion-mediated thrombocytopenia. Prognosis includes treatment with fresh frozen plasma (replaces the coagulation factors) and platelets (Hardy et al., 2005). Frequent platelet concentrate transfusions have shown to ameliorate the survival of such patients (Cosgriff et al., 1997; Cinat et al., 1999).

OTHER CAUSES OF THROMBOCYTOPENIA

Cyclic Thrombocytopenia

Cyclic thrombocytopenia (CT) is a rare disorder. It is described by the periodic fluctuations in platelet count, with episodes of thrombocytopenia, followed by gradual elevation in platelet count (Jose et al., 2016). The exact etiology of CT is unclear. Platelet count is inversely proportional to the fluctuating levels of TPO (Yujiri et al., 2009). One of the explanations for the cyclic fluctuation can be the variation in the macrophage F_c receptor expression to the changing hormones (Rice et al., 2001).

Rebound thrombocytosis is characteristic of CT. Most of the CT cases are idiopathic and hence, can be easily confused with ITP. Nevertheless, few cases are also reported in association with myeloproliferative neoplasms (Abe et al., 2000; Steensma et al., 2001). Autoimmune platelet destruction, megakaryocytic, hormonal disturbances, infections, and hypoplasia/aplasia can be the probable mechanisms in CT.

The bleeding tendency in CT ranges from asymptomatic to easy bruising (gingival bleeding, menorrhagia, recurrent epistaxis, and hematuria), to severe bleeding (gastrointestinal or cerebral hemorrhage) (Go, 2005). Patients diagnosed for ITP, but unresponsive to treatments (such as IVIg, glucocorticoids or splenectomy) must be considered for CT. Patients with CT are found to be responsive to cyclosporine and hormone therapy. CT is found to be common in females, and oral contraceptives are recommended to delay the menstrual cycle and cover the thrombocytopenic cycle. Bleeding symptoms can also be reduced with the use of antifibrinolytic drugs (tranexamic acid or aminocaproic acid).

ANTIOXIDANTS AS THERAPEUTICS IN THROMBOCYTOPENIA

Current management and therapy for thrombocytopenia include pharmacological and surgical treatments, and platelet transfusions under severe conditions. However, the unavailability of supportive medication to increase the platelet count is often fatal to patients (Arollado et al., 2013). Therefore, there is a need to explore the various antioxidants and plant extracts for their anti-thrombocytopenic potential.

The plant extracts of *Carica papaya* (Arollado et al., 2013; Subenthiran et al., 2013) *Euphorbia hirta* (Arollado et al., 2013; Apostol et al., 2012) *Ipomea batatas, Althernanthera sessilis* and *Momordica charantia* (Arollado et al., 2013) have shown to improve the platelet numbers during thrombocytopenia. Aqueous leaf extract of *C. papaya* has been extensively studied for its platelet enhancing effect during dengue infection and dengue hemorrhagic fever. Polyphenols contained in *C. papaya* leaf extract possess an antioxidant potential and contribute to its platelet augmenting property (Subenthiran et al., 2013; Kasture et al., 2016).

The antioxidants such as ascorbic acid (Brox et al., 1988; Jubelirer, 1993), vitamin E and vitamin C (Chandra et al., 2013) have demonstrated their ability to increase platelet count in ITP and early recovery of dengue patients. Accordingly, the anti-thrombocytopenic properties of other antioxidants need to be explored.

CONCLUSION

Platelets play a significant role in repairing the damaged endothelium of blood vessels. They arrest bleeding at the site of vascular injury through a process involving platelet adhesion, activation, secretion, and aggregation, subsequently leading to the formation of a hemostatic plug. Disorders in platelets (numbers, functions, or morphology) can lead to various bleeding disorders. Thrombocytopenia is a platelet disorder caused due to the reduction in platelet number ($< 150 \times 10^9$ cells/ L) and is further classified based on the different pathophysiology. Various causative factors form the basis for the strategies employed in thrombocytopenia treatments. However, the correct diagnosis of the type of thrombocytopenia is crucial to avoid complications during treatment. Recent studies have shown the efficacy of various antioxidants and plant extracts in treating some of the thrombocytopenic conditions. Nevertheless further research is essential in this field in order to progress towards reliable therapeutics with minimal side effects.

REFERENCES

[1] Abe, Y., Hirase, N., Muta, K., Okada, Y., Kimura, T., Umemura, T., Nishimura, J. & Nawata, H. (2000). "Adult onset cyclic hematopoiesis in a patient with myelodysplastic syndrome." *International Journal of Hematology, 71,* 40-45.

[2] Alamelu, J. & Liesner, R. (2010). "Modern management of severe platelet function disorders." *British Journal of Haematology, 149,* 813-823.

[3] Angotti, L. B., Post, G. R., Robinson, N. S., Lewis, J. A., Hudspeth, M. P. & Lazarchick, J. (2008). "Pancytopenia with myelodysplasia due to copper deficiency." *Pediatric Blood and Cancer, 51*, 693-695.

[4] Apostol, J. G., Gan, J. V. A., Raynes, R. J. B., Sabado, A. A. S., Carigma, A. Q., Santiago, L. A. & Ysrael, M. C. (2012). "Platelet-increasing effects of *Euphorbia hirta* Linn. (Euphorbiaceae) in ethanol-induced thrombocytopenia rat models. *International Journal of Pharmaceutical Frontier Research, 2*, 1-11.

[5] Arnal, C., Piette, J. C., Leone, J., Taillan, B., Hachulla, E., Roudot-Thoraval, F., Papo, T., Schaeffer, A., Bierling, P. & Godeau, B. (2002). "Treatment of severe immune thrombocytopenia associated with systemic lupus erythematosus: 59 cases." *The Journal of Rheumatology, 29*, 75-83.

[6] Arnold, D. M., Kukaswadia, S., Nazi, I., Esmail, A., Dewar, L., Smith, J. W., Warkentin, T. E. & Kelton, J. G. (2013). "A systematic evaluation of laboratory testing for drug-induced immune thrombocytopenia." *Journal of Thrombosis and Haemostasis, 11*, 169-176.

[7] Arollado, E. C., Pena, I. G. & Dahilig, V. R. A. (2013). "Platelet augmentation activity of selected Philippine plants." *International Journal of Pharmaceutical and Phytopharmacological Research, 3*, 121-123.

[8] Aster, R. H. (1966). "Pooling of platelets in the spleen: Role in the pathogenesis of "hypersplenic" thrombocytopenia." *Journal of Clinical Investigation, 45*, 645-657.

[9] Aster, R. H. (1972). "Platelet sequestration studies in man." *British Journal of Haematology, 22*, 259-263.

[10] Aster, R. H. & Bougie, D. W. (2007). "Drug-induced immune thrombocytopenia." *The New England Journal of Medicine, 357*, 580-587.

[11] Bakchoul, T. & Greinacher, A. (2016). *Molecular and cellular biology of platelet formation: Implications in health and disease.* Switzerland: Springer International Publishing.

[12] Balduini, C. L., Pecci, A. & Noris, P. (2015). "Diagnosis and management of inherited thrombocytopenias." *Seminars in Thrombosis and Hemostasis, 39*, 161-171.

[13] Balduini, C. L., Pecci, A. & Savoia, A. (2011). "Recent advances in the understanding and management of MYH9-inherited thrombocytopenia." *British Journal of Haematology, 154*, 161-174.

[14] Ballard, H. S. (2015). "The hematological complications of alcoholism." *Alcohol Health and Research World, 21*, 45-52.

[15] Berkman, N., Michaeli, Y., Or, R. & Eldor, A. (1991). "EDTA-dependent pseudothrombocytopenia: A clinical study of 18 patients and a review of the literature." *American Journal of Hematology, 36*, 195-201.

[16] Berns, J. S., Kaplan, B. S., Mackow, R. C. & Hefter, L. G. (1992). "Inherited hemolytic uremic syndrome in adults." *American Journal of Kidney Diseases, 19*, 331-334.

[17] Bessman, J. D. (1984). "The relation of megakaryocyte ploidy to platelet volume." *American Journal of Hematology, 16*, 161-170.

[18] Bianchi, V., Robles, R., Alberio, L., Furlan, M. & Lammle, B. (2002). "von Willebrand factor cleaving protease (ADAMTS13) in thrombocytopenic disorders: A severely deficient activity is specific for thrombotic thrombocytopenic purpura." *Blood, 100*, 710-713.

[19] Bizzaro, N. (1995). "EDTA-dependent pseudothrombocytopenia: A clinical and epidemiological study of 112 cases, with 10-year follow-up." *American Journal of Hematology, 50*, 103-109.

[20] Bolton-Maggs, P. H., Chalmers, E. A., Collins, P. W., Harrison, P., Kitchen, S., Liesner, R. J., Minford, A., Mumford, A. D., Parapia, L. A., Perry, D. J., Watson, S. P., Wilde, J. T., Williams, M. D. & U. K. H. C. D. O. (2006). A review of inherited platelet disorders with guidelines for their management on behalf of the UKHCDO." *British Journal of Haematology, 135*, 603-633.

[21] Boumpas, D. T., Barez, S., Klippel, J. H. & Balow, J. E. (1990). "Intermittent cyclophosphamide for the treatment of autoimmune thrombocytopenia in systemic lupus erythematosus. *Annals of Internal Medicine, 112*, 674-677.

[22] Brox, A. G., Howson-Jan, K. & Fauser, A. A. (1988). "Treatment of idiopathic thrombocytopenic purpura with ascorbate." *British Journal of Haematology*, *70*, 341-344.

[23] Bunimov, N., Fuller, N. & Hayward, C. P. (2013). "Genetic loci associated with platelet traits and platelet disorders." *Seminars in Thrombosis and Hemostasis*, *39*, 291-305.

[24] Burns, E. R. & Lawrence, C. (1989). "Bleeding time. A guide to its diagnostic and clinical utility." *Archives of Pathology and Laboratory Medicine*, *113*, 1219-1224.

[25] Casonato, A., Bertomoro, A., Pontara, E., Dannhauser, D., Lazzaro, A. R. & Girolami, A. (1994). "EDTA dependent pseudo-thrombocytopenia caused by antibodies against the cytoadhesive receptor of platelet GPIIB-IIIA." *Journal of Clinical Pathology*, *47*, 625-630.

[26] Chakravorty, S. & Roberts, I. (2011). "How I manage neonatal thrombocytopenia." *British Journal of Haematology*, *156*, 155-162.

[27] Chandra, P., Sharma, H., Gupta, A. & Rai, Y. (2013). "Role of antioxidant vitamin E and C on platelet levels in dengue fever." *International Journal of Applied and Basic Medical Research*, *3*, 287-291.

[28] Christie, D. J., Mullen, P. C. & Aster, R. H. (1985). "Fab-mediated binding of drug-dependent antibodies to platelets in quinidine- and quinine-induced thrombocytopenia." *Journal of Clinical Investigation*, *75*, 310-314.

[29] Cinat, M. E., Wallace, W. C., Nastanski, F., West, J., Sloan, S., Ocariz, J. & Wilson, S. E. (1999). "Improved survival following massive transfusion in patients who have undergone trauma." *Archives of Surgery*, *134*, 964-968.

[30] Cines, D. B. & Blanchette, V. S. (2002). "Immune thrombocytopenic purpura." *The New England Journal of Medicine*, *346*, 995-1008.

[31] Cohen, M. G. & Li, E. K. (1991). "Limited effects of intravenous IgG in treating systemic lupus erythematosus-associated thrombocytopenia." *Arthritis and Rheumatism*, *34*, 787-788.

[32] Cole, J. L., Marzec, U. M., Gunthel, C. J., Karpatkin, S., Worford, L., Sundell, I. B., Lennox, J. L., Nichol, J. L. & Harker, L. A. (1998). "Ineffective platelet production in thrombocytopenic human immunodeficiency virus-infected patients." *Blood*, *91*, 3239-3246.

[33] Cosgriff, N., Moore, E. E., Sauaia, A., Kenny-Moynihan, M., Burch, J. M. & Galloway, B. (1997). "Predicting life-threatening coagulopathy in the massively transfused trauma patient: Hypothermia and acidoses revisited." *The Journal of Trauma*, *42*, 857-861.

[34] Cuker, A. (2010). "Toxicities of the thrombopoietic growth factors." *Seminars in Hematology*, *47*, 289-298.

[35] Curtis, B. R., McFarland, J. G., Wu, G. G., Visentin, G. P. & Aster, R. H. (1994). "Antibodies in sulfonamide- induced immune thrombocytopenia recognize calcium-dependent epitopes on the glycoprotein IIb/IIIa complex." *Blood*, *84*, 176-183.

[36] De Paepe, A. & Malfait, F. (2004). "Bleeding and bruising in patients with Ehlers-Danlos syndrome and other collagen vascular disorders." *British Journal of Haematology*, *127*, 491-500.

[37] Ding, C., Foote, S. & Jones, G. (2008). "B-cell-targeted therapy for systemic lupus erythematosus: An update." *Bio Drugs*, *22*, 239-249.

[38] Diz-Kucukkaya, R. & Lopez, J. A. (2013). "Inherited disorders of platelets: Membrane glycoprotein disorders." *Hematology/Oncology Clinics of North America*, *27*, 613-627.

[39] Dominguez, A., Gamallo, G., Garcia, R., Lopez-Pastor, A., Pena, J. M. & Vazquez, J. J. (1994). "Pathophysiology of HIV related thrombocytopenia: An analysis of 41 patients." *Journal of Clinical Pathology*, *47*, 999-1003.

[40] Drachman, J. G. (2004). "Inherited thrombocytopenia: when a low platelet count does not mean ITP." *Blood*, *103*, 39-398.

[41] Dubansky, A. S., Boyett, J. M., Falletta, J., Mahoney, D. H., Land, V. J., Pullen, J. & Buchanan, G. (1989). "Isolated thrombocytopenia in children with acute lymphoblastic leukemia: a rare event in a Pediatric Oncology Group Study." *Pediatrics*, *84*, 1068-1071.

[42] Easterbrook, P. J. & Davis, H. P. (1985). "Thrombocytopenia in hypothermia: A common but poorly recognised complication." *British Medical Journal (Clinical Research Ed), 291*, 23.

[43] Essein, F. A., Ojeda, H. F., Salameh, J. R., Baker, K. R., Rice, L. & Sweeney, J. F. (2003). "Laparoscopic splenectomy for chronic recurrent thrombotic thrombocytopenic purpura." *Surgical Laparoscopy Endoscopy & Percutaneous Techniques, 13*, 218-221.

[44] Fang, C. H., Chien, Y. L., Yang, L. M., Lu, W. J. & Lin, M. F. (2015). "EDTA-dependent pseudothrombocytopenia." *Formosan Journal of Surgery, 48*, 107-109.

[45] Favaloro, E. J. (2001). "Utility of the PFA-100 for assessing bleeding disorders and monitoring therapy: A review of analytical variables, benefits and limitations." *Haemophilia, 7*, 170-179.

[46] Filipovich, A. H., Stone, J. V., Tomany, S. C., Ireland, M., Kollman, C., Pelz, C. J., Casper, J. T., Cowan, M. J., Edwards, J. R., Fasth, A., Gale, R. P., Junker, A., Kamani, N. R., Loechelt, B. J., Pietryga, D. W., Ringden, O., Vowels, M., Hegland, J., Williams, A. V., Klein, J. P., Sobocinski, K. A., Rowlings, P. A. & Horowitz, M. M. (2001). "Impact of donor type on outcome of bone marrow transplantation for Wiskott–Aldrich syndrome: collaborative study of the International Bone Marrow Transplant Registry and the National Marrow Donor Program." *Blood, 97*, 1598-1603.

[47] Flaujac, C., Boukour, S. & Cramer-Borde, E. (2010). "Platelets and viruses: an ambivalent relationship." *Cellular and Molecular Life Sciences, 67*, 545-556.

[48] Font, J., Jimenez, S., Cervera, R., Garcia-Carrasco, M., Ramos-Casals, M., Campdelacreu, J. & Ingelmo, M. (2000). "Splenectomy for refractory Evans' syndrome associated with antiphospholipid antibodies: Report of two cases." *Annals of the Rheumatic Diseases, 59*, 920-923.

[49] Froom, P. & Barak, M. (2011). "Prevalence and course of pseudothrombocytopenia in outpatients." *Clinical Chemistry and Laboratory Medicine, 49*, 111-114.

[50] Furlan, M. & Lammle, B. (2001). "Aetiology and pathogenesis of thrombotic thrombocytopenic purpura and haemolytic uremic syndrome: the role of von Willebrand factor cleaving protease." *Best Practice and Research: Clinical Haematology, 14*, 437-454.

[51] Furlan, M., Robles, R., Galbusera, M., Remuzzi, G., Kyrle, P. A., Brenner, B., Krause, M., Scharrer, I., Aumann, V., Mittler, U., Solenthaler, M. & Lammle, B. (1998). "von Willebrand factor-cleaving protease in thrombotic thrombocytopenic purpura and the hemolytic-uremic syndrome." *The New England Journal of Medicine 339*, 1578-1584.

[52] Galli, M., Daldossi, M. & Barbui, T. (1994). "Anti-glycoprotein Ib/IX and IIb/IIIa antibodies in patients with antiphospholipid antibodies." *Thrombosis and Haemostasis, 71*, 571-575.

[53] Galli, M., Finazzi, G. & Barbu, T. (1996). "Thrombocytopenia in the antiphospholipid syndrome." *British Journal of Haematology, 93*, 1-5.

[54] Gando, S., Wada, H., Thachil, J. & Scientific and Standardization Committee on DIC of the International Society on Thrombosis and Haemostasis (ISTH). (2013) "Differentiating disseminated intravascular coagulation (DIC) with the fibrinolytic phenotype from coagulopathy of trauma and acute coagulopathy of trauma-shock (COT/ACOTS)." *Journal of Thrombosis and Haemostasis, 11*, 826-835.

[55] Garcia Suarez, J., Merino, J. L., Rodríguez, M., Velasco, A. & Moreno, M. C. (1991). "Pseudothrombocytopenia: Incidence, causes and methods of detection." *Sangre (Barc), 36*, 197-200.

[56] Gauer, R. L. & Braun, M. M. (2012). "Thrombocytopenia." *American Family Physician, 85*, 612-622.

[57] George, J. N. (2000). "How I treat patients with thrombotic thrombocytopenic purpurahemolytic uremic syndrome." *Blood, 96*, 1223-1229.

[58] George, J. N. & Saucerman, S. (1988). "Platelet IgG, IgA, IgM, and albumin: Correlation of platelet and plasma concentrations in normal subjects and in patients with ITP or dysproteinemia." *Blood, 72*, 362-365.

[59] Go, R. S. (2005). "Idiopathic cyclic thrombocytopenia." *Blood Reviews*, *19*, 53-59.

[60] Gordon, N., Thom, J., Cole, C. & Baker, R. (1995). "Rapid detection of hereditary and acquired platelet storage pool deficiency by flow cytometry." *British Journal of Haematology*, *89*, 117-123.

[61] Gowland, E., Kay, H. E., Spillman, J. C. & Williamson, J. R. (1969). "Agglutination of platelets by a serum factor in the presence of EDTA." *Journal of Clinical Pathology*, *22*, 460-464.

[62] Green, P. (2012). "Anemias beyond B12 and iron deficiency: The buzz about other B's elementary, and nonelementary problems." *Hematology. American Society of Hematology. Education Program*, 2012, 492-498.

[63] Greinacher, A., Friesecke, S., Abel, P., Dressel, A., Stracke, S., Fiene, M., Ernst, F., Selleng, K., Weissenborn, K., Schmidt, B. M., Schiffer, M., Felix, S. B., Lerch, M. M., Kielstein, J. T. & Mayerle, J. (2011). "Treatment of severe neurological deficits with IgG depletion through immunoadsorption in patients with *Escherichia coli* O104:H4-associated haemolytic uraemic syndrome: a prospective trial." *Lancet*, *378*, 1166-1173.

[64] Gresele, P. (2015). "Diagnosis of inherited platelet function disorders: Guidance from the SSC of the ISTH." *Journal of Thrombosis and Haemostasis*, *13*, 314-322.

[65] Hardy, J. F., de Moerloose, P. & Samama, C. M. (2005). "The coagulopathy of massive transfusion." *Vox Sanguinis*, *89*, 123-127.

[66] Harrison, P. (2005). "Platelet function analysis." *Blood Reviews*, *19*, 111-23.

[67] Hellstrom-Lindberg, E., Kanter-Lewensohn, L., Nichol, J. & Ost, A. (1999). "Spontaneous and cytokine-induced thrombocytopenia in myelodysplastic syndromes: Serum thrombopoietin levels and bone marrow morphology. Scandinavian MDS Group, Sweden and Norway." *British Journal of Haematology*, *105*, 966-973.

[68] Hochberg, M. C. (1997). "Updating the American College of Rheumatology revised criteria for the classification of systemic lupus erythematosus." *Arthritis and Rheumatism*, *40*, 1725.

[69] Hoffmeister, K. M., Felbinger, T. W., Falet, H., Denis, C. V., Bergmeier, W., Mayadas, T. N., von Andrian, U. H., Wagner, D. D., Stossel, T. P. & Hartwig, J. H. (2003). "The clearance mechanism of chilled blood platelets." *Cell, 112*, 87-97.

[70] Isik, A., Balcik, O. S., Akdeniz, D., Cipil, H., Uysal, S. & Kosar, A. (2012). "Relationship between some clinical situations, auto-antibodies, and pseudothrombocytopenia." *Clinical and Applied Thrombosis/Hemostasis, 18*, 645-649.

[71] Jandl, J. H., Aster, R. H., Forkner, C. E., Fisher, A. M. & Vilter, R. W. (1967). "Splenic pooling and the pathophysiology of hypersplenism." *Transactions of the American Clinical and Climatological Association, 78*, 9-27.

[72] Jilma, B. (2001). "Platelet function analyzer (PFA-100): A tool to quantify congenital or acquired platelet dysfunction." *Journal of Laboratory and Clinical Medicine, 138*, 152-163.

[73] Johns, C. S. (2004). "Platelet function testing." *Clinical Hemostasis Review, 18*, 1-8.

[74] Joly, B. S., Coppo, P. & Veyradier, A. (2017). "Thrombotic thrombocytopenic purpura." *Blood, 129*, 2836-2846.

[75] Jose, N., Karthik, G., Jasmine, S. & Sathyendra, S. (2016). "An Unusual Cause of Recurrent Thrombocytopenia." *Journal of Case Reports, 6*, 188-190.

[76] Jubelirer, S. J. (1993). "Pilot study of ascorbic acid for the treatment of refractory immune thrombocytopenic purpura." *American Journal of Hematology, 43*, 44-46.

[77] Kamath, V., Sarda, P., Chacko, M. P. & Sitaram, U. (2013). "Pseudothrombocytopenia observed with ethylene diamine tetra acetate and citrate anticoagulants, resolved using 37°C incubation and Kanamycin." *Indian Journal of Pathology and Microbiology, 56*, 306-308.

[78] Kaplan, C. (1998). "Evaluation of serological platelet antibody assays." *Vox Sanguinis, 74*, 355-358.

[79] Kasper, D. L., Braunwald, E., Fauci, A. S., Hauser, S. L., Longo, D. L. & Jameson, J. L. *Harrison's principles of internal medicine*. New York: McGraw Hill Medical.

[80] Kasture, P. N., Nagabhushan, K. H. & Kumar, A. (2016). "A multicentric, double-blind, placebo-controlled, randomized. Prospective study to evaluate the efficacy and safety of *Carica papaya* leaf extract, as empirical therapy for thrombocytopenia associated with dengue fever." *Journal of the Association of Physicians of India*, *64*, 15-20.

[81] Kaushansky, K., Lichtman, M. A., Prchal, J. T., Levi, M., Press, O. W., Burns, L. J. & Caliguiri, M. A. (2016). *Williams Hematology*. New York: McGraw-Hill Education.

[82] Kelton, J. G., Powers, P. J. & Carter, C. J. (1982). "A prospective study of the usefulness of the measurement of platelet-associated IgG for the diagnosis of idiopathic thrombocytopenic purpura." *Blood, 60*, 1050-1053.

[83] Kiefel, V., Santoso, S., Weisheit, M. & Mueller-Eckhardt, C. (1987). "Monoclonal antibody specific immobilization of platelet antigens (MAIPA): A new tool for the identification of platelet-reactive antibodies." *Blood, 70*, 1722-1726.

[84] Kim, M. J., Park, P. W., Seo, Y. H., Kim, K. H., Seo, J. Y., Jeong, J. H., Park, M. J. & Ahn, J. Y. (2014). "Comparison of platelet parameters in thrombocytopenic patients associated with acute myeloid leukemia and primary immune thrombocytopenia." *Blood Coagulation and Fibrinolysis, 25*, 221-225.

[85] Kobayashi, T., Terao, T., Maki, M. & Ikenoue, T. (2001). "Diagnosis and management of acute obstetrical DIC." *Seminars in Thrombosis and Hemostasis, 27*, 161-167.

[86] Lapeyraque, A. L., Malina, M., Fremeaux-Bacchi, V., Boppel, T., Kirschfink, M., Oualha, M., Proulx, F., Clermont, M. J., Le Deist, F., Niaudet, P. & Schaefer, F. (2011). "Eculizumab in severe Shiga-toxin-associated HUS." *The New England Journal of Medicine, 364*, 2561-2563.

[87] Latvala, J., Parkkila, S. & Niemela, O. (2004). "Excess alcohol consumption is common in patients with cytopenia: Studies in blood

and bone marrow cells." *Alcoholism: Clinical and Experimental Research, 28,* 619-624.

[88] Lawrence, S. P., Lezotte, D. C., Durham, J. D., Kumpe, D. A., Everson, G. T. & Bilir, B. M. (1995). "Course of thrombocytopenia of chronic liver disease after transjugular intrahepatic portosystemic shunts (TIPS). A retrospective analysis." *Digestive Diseases and Sciences, 40,* 1575-1580.

[89] Leach, M. F., Cooper, L. K. & AuBuchon, J. P. (1997). "Detection of drug-dependent, platelet-reactive antibodies by solid-phase red cell adherence assays." *British Journal of Haematology, 97,* 755-761.

[90] Letsky, E. A. & Greaves, M. (1996). "Guidelines on the investigation and management of thrombocytopenia in pregnancy and neonatal alloimmune thrombocytopenia. Maternal and Neonatal Haemostasis Working Party of the Haemostasis and Thrombosis Task Force of the British Society for Haematology." *British Journal of Haematology, 95,* 21-26.

[91] Li, S., Wang, L., Zhao, C., Li, L., Peng, J. & Hou, M. (2007). "CD8+ T cells suppress autologous megakaryocyte apoptosis in idiopathic thrombocytopenic purpura." *British Journal Haematology, 139,* 605-611.

[92] Lindenbaum, J. & Lieber, C. S. (1969). "Hematologic effects of alcohol in man in the absence of nutritional deficiency." *The New England Journal of Medicine, 281,* 333-338.

[93] Liu, B., Zhao, H. & Poon, M. D. (2007). "Abnormality of CD4(+)CD25(+) regulatory T cells in idiopathic thrombocytopenic purpura." *European Journal of Haematology, 78,* 139-143.

[94] Lombarts, A. J. & de Kieviet, W. (1988). "Recognition and prevention of pseudothrombocytopenia and concomitant pseudoleukocytosis." *American Journal of Clinical Pathology, 89,* 634-639.

[95] Lopez, J. A., Li, C. Q., Weisman, S. & Chambers, M. (1995). "The glycoprotein Ib-IX complex-specific monoclonal antibody SZ1 binds to a conformation-sensitive epitope on glycoprotein IX: Implications for the target antigen of quinine/quinidine-dependent autoantibodies." *Blood, 85,* 1254-1258.

[96] Louache, F. & Vainchenker, W. (1994). "Thrombocytopenia in HIV infection." *Current Opinion in Hematology*, *1*, 369-372.

[97] Lucas, G. F., Pittman, S. J., Davies, S., Solanki, T. & Bruggemann, K. (1997). "Post-transfusion purpura (PTP) associated with anti-HPA-1a, anti-HPA-2b and anti-HPA-3a antibodies." *Transfusion Medicine*, *7*, 295-299.

[98] Lusher, J. M. & Iyer, R. (1977). "Idiopathic thrombocytopenic purpura in children." *Seminars in Thrombosis and Hemostasis*, *3*, 175-199.

[99] Maier, W. P., Gordon, D. S., Howard, R. F., Saleh, M. N., Miller, S. B., Lieberman, J. D. & Woodlee, P. M. (1990). "Intravenous immunoglobulin therapy in systemic lupus erythematosus-associated thrombocytopenia." *Arthritis and Rheumatism*, *33*, 1233-1239.

[100] Maslanka, K., Marciniak-Bielak, D. & Szczepinski, A. "Pseudo-thrombocytopenia in blood donors." *Vox Sanguinis*, *95*, 349.

[101] Massey, G. V., McWilliams, N. B., Mueller, D. G., Napolitano, A. & Maurer, H. M. (1987). "Intravenous immunoglobulin in treatment of neonatal isoimmune thrombocytopenia." *The Journal of Pediatrics*, *111*, 133-135.

[102] McCormick, P. A. & Murphy, K. M. (2000). "Splenomegaly, hypersplenism and coagulation abnormalities in liver disease." *Bailliere's Best Practice and Research: Clinical Gastroenterology*, *14*, 1009-1031.

[103] McKenzie, C. G., Guo, L., Freedman, J. & Semple, J. W. (2013). "Cellular immune dysfunction in immune thrombocytopenia (ITP)." *British Journal of Haematology*, *163*, 10-23.

[104] McMillan, R. (2000). "Autoantibodies and autoantigens in chronic immune thrombocytopenic purpura." *Seminars in Hematology*, *37*, 239-248.

[105] McMorran, B. J., Wieczorski, L., Drysdale, K. E., Chan, J. A., Huang, H. M., Smith, C., Mitiku, C., Beeson, J. G., Burgio, G. & Foote, S. J. (2012). "Platelet factor 4 and Duffy antigen required for platelet killing of *Plasmodium falciparum*." *Science*, *338*, 1348-1351.

[106] Michel, M., Lee, K., Piette, J. C., Fromont, P., Schaeffer, A., Bierling, P. & Godeau, B. (2002). "Platelet autoantibodies and lupus-associated thrombocytopenia." *British Journal of Haematology*, *119*, 354-358.
[107] Michelson, A. D. (2007). *Platelets*. San Diego, CA: Academic Press.
[108] Michot, F. & Gut, J. (1987). "Alcohol-induced bone marrow damage. A bone marrow study in alcohol dependent individuals." *Acta Haematologica*, *78*, 252-257.
[109] Minchinton, R. M., Cunningham, I., Cole-Sinclair, M., Van der Weyden, M., Vaughan, S. & McGrath, K. M. (1990). "Autoreactive platelet antibody in post transfusion purpura." *Australia and New Zealand Journal of Medicine*, *20*, 111-115.
[110] Miyakis, S., Lockshin, M. D., Atsumi, T., Branch, D. W., Brey, R. L., Cervera, R., Derksen, R. H., DE Groot, P. G., Koike, T., Meroni, P. L., Reber, G., Shoenfeld, Y., Tincani, A., Vlachoyiannopoulos, P. G. & Krilis, S. A. (2006). "International consensus statement on an update of the classification criteria for definite antiphospholipid syndrome (APS)." *Journal of Thrombosis and Haemostasis*, *4*, 295-306.
[111] Murphy, M. F., Verjee, S. & Greaves, M. (1999). "Inadequacies in the postnatal management of fetomaternal alloimmune thrombocytopenia (FMAIT)." *British Journal of Haematology*, *105*, 123-126.
[112] Nahirniak, S., Slichter, S. J., Tanael, S., Rebulla, P., Pavenski, K., Vassallo, R., Fung, M., Duquesnoy, R., Saw, C. L., Stanworth, S., Tinmouth, A., Hume, H., Ponnampalam, A., Moltzan, C., Berry, B., Shehata, N. G. & International Collaboration for Transfusion Medicine. (2015). "Guidance on platelet transfusion for patients with hypoproliferative thrombocytopenia." *Transfusion Medicine Reviews*, *29*, 3-13.
[113] Neild, G. (1987). "The haemolytic uraemic syndrome: A review." *The Quarterly Journal of Medicine*, *241*, 367-376.
[114] Neunert, C., Lim, W., Crowther, M., Cohen, A., Solberg, L. Jr., Crowther, M. A. & American Society of Hematology. 2011. "The American Society of Hematology. (2011). evidence-based practice guideline for immune thrombocytopenia." *Blood*, *117*, 4190-4207.

[115] Nicol, K. K., Shelton, B. J., Knovich, M. S. & Owen, J. (2003). "Overweight individuals are at increased risk for thrombotic thrombocytopenic purpura." *American Journal of Hematology, 74,* 170-174.

[116] Nieminen, U. & Kekomaki, R. (1992). "Quinidine-induced thrombocytopenic purpura: Clinical presentation in relation to drug-dependent and drug-independent platelet antibodies." *British Journal of Haematology, 80,* 77-82.

[117] Nithish, B., Vikram, G. & Hariprasad, S. (2011). "Thrombocytopenia in malaria: A clinical study." *Biomedical Research, 22,* 489-491.

[118] Olariu, M., Olariu, C. & Olteanu, D. (2010). "Thrombocytopenia in chronic hepatitis C." *Journal of Gastrointestinal and Liver Diseases, 19,* 381-385.

[119] Payne, B. A. & Pierre, R. V. (1984). "Pseudothrombocytopenia: a laboratory artifact with potentially serious consequences." *Mayo Clinic Proceedings, 59,* 123-125.

[120] Pegels, J. G., Bruynes, E. C., Engelfriet, C. P. & von dem Borne, A. E. (1982). "Pseudothrombocytopenia: an immunologic study on platelet antibodies dependent on ethylene diamine tetra-acetate." *Blood, 59,* 157-161.

[121] Peterson, P., Hayes, T. E., Arkin, C. F., Bovill, E. G., Fairweather, R. B., Rock, Jr. W. A., Triplett, D. A. & Brandt, J. T. (1998). "The preoperative bleeding time test lacks clinical benefit: College of American Pathologists' and American Society of Clinical Pathologists' position article." *Archives of Surgery, 133,* 134-139.

[122] Pina-Cabral, J. M., Ribeiro-da-Silva, A. & Almeida-Dias, A. (1985). "Platelet sequestration during hypothermia in dogs treated with sulphinpyrazone and ticlopidine - reversibility accelerated after intra-abdominal rewarming." *Thrombosis and Haemostasis, 54,* 838-841.

[123] Pujol, M., Ribera, A., Vilardell, M., Ordi, J. & Feliu, E. (1995). "High prevalence of platelet autoantibodies in patients with systemic lupus erythematosus." *British Journal of Haematology, 89,* 137-141.

[124] Rabinowitz, Y. & Dameshek, W. (1960). "Systemic lupus erythematosus after "idiopathic" thrombocytopenic purpura: A review." *Annals of Internal Medicine*, *52*, 1-28.

[125] Rice, L., Nichol, J. L., McMillan, R., Roskos, L. K. & Bacile, M. 2001. "Cyclic immune thrombocytopenia responsive to thrombopoietic growth factor therapy." *American Journal of Hematology*, *68*, 210-214.

[126] Richey, M. E., Gilstrap, L. C. 3rd & Ramin, S. M. (1995). "Management of disseminated intravascular coagulopathy." *Clinical Obstetrics and Gynecology*, *38*, 514-520.

[127] Ridolfi, R. L. & Bell, W. R. (1981). "Thrombotic thrombocytopenia purpura: Report of 25 cases and a review of the literature." *Medicine*, *60*, 413-428.

[128] Ruggenenti, P., Noris, M. & Remuzzi, G. (2001). "Thrombotic microangiopathy, haemolytic uremic syndrome and thrombotic thrombocytopenic purpura." *Kidney International*, *60*, 831-846.

[129] Saito, H. (1993). "Thrombotic microangiopathy (DIC, TTP and HUS) - recent advances in pathogenesis and management." *Nihon Rinsho*, *51*, 5-10.

[130] Sakurai, S., Shiojima, I., Tanigawa, T. & Nakahara, K. (1997). "Aminoglycosides prevent and dissociate the aggregation of platelets in patients with EDTA-dependent pseudothrombocytopenia." *British Journal of Haematology*, *99*, 817-823.

[131] Santoso, S. & Kiefel, V. (2001). "Human platelet alloantigens." *Wiener klinische Wochenschrift*, *113*, 806-813.

[132] Scully, M. (2012). "Rituximab in the treatment of TTP." *Hematology*, *17*, S22-S24.

[133] Scully, M., Hunt, B. J., Benjamin, S., Liesner, R., Rose, P., Peyvandi, F., Cheung, B., Machin, S. J. & British Committee for Standards in Haematology. (2012). "Guidelines on the diagnosis and management of thrombotic thrombocytopenic purpura and other thrombotic microangiopathies." *British Journal of Haematology*, *158*, 323-335.

[134] Seligsohn, U. (2012). "Treatment of inherited platelet disorders." *Haemophilia*, *18*, 161-165.

[135] Shreiner, D. P. & Bell, W. R. (1973). "Pseudothrombocytopenia: manifestations of a few type of platelet agglutinin." *Blood*, *42*, 541-549.

[136] Shulman, N. R., Weinrach, R. S., Libre, E. P. & Andrews, H. L. 1965. "The role of the reticuloendothelial system in the pathogenesis of idiopathic thrombocytopenic purpura." *Transactions of the Association of American Physicians*, *78*, 374-390.

[137] Sihler, K. C. & Napolitano, L. M. (2009). "Massive transfusion: New insights." *Chest*, *136*, 1654-1667.

[138] Smith, C. M., Tobin, J. D. Jr. Burris, S. M. & White, J. G. (1992). "Alcohol consumption in the guinea pig is associated with reduced megakaryocyte deformability and platelet size." *Journal of Laboratory and Clinical Medicine.*, *120*, 699-706.

[139] Spencer, J. A. & Burrows, R. F. (2001). "Feto-maternal alloimmune thrombocytopenia: A literature review and statistical analysis." *Australian and New Zealand Journal of Obstetrics and Gynaecology*, *41*, 45-55.

[140] Stasi, R. (2001). "Therapeutic strategies for hepatitis- and other infection-related immune thrombocytopenias." *Seminars in Hematology*, *46*, S15-S25.

[141] Stasi, R., Pagano, A., Stipa, E. & Amadori, S. (2001). "Rituximab chimeric anti-CD20 monoclonal antibody treatment for adults with chronic idiopathic thrombocytopenic purpura." *Blood*, *98*, 952-957, 2001.

[142] Stasi, R., Stipa, E., Masi, M., Cecconi, M., Scimo, M. T., Oliva, F., Sciarra, A., Perrotti, A. P., Adomo, G., Amadori, S. & Papa G. (1995). "Long-term observation of 208 adults with chronic idiopathic thrombocytopenic purpura." *The American Journal of Medicine*, *98*, 436-442.

[143] Steensma, D. P., Harrison, C. N. & Tefferi, A. (2001). "Hydroxyurea-associated platelet count oscillations in polycythemia vera: A report of four new cases and a review." *Leukemia and Lymphoma*, *42*, 1243-1253.

[144] Subenthiran, S., Choon, T. C., Cheong, K. C., Thayan, R., Teck, M. B., Muniandy, P. K., Afzan, A., Abdullah, N. R. & Ismail, Z. (2013). "*Carica papaya* leaves juice significantly accelerates the rate of increase in platelet count among patients with dengue fever and dengue haemorrhagic fever." *Evidence Based Complementary and Alternative Medicine* Article ID 616737, 7 pages.

[145] Sullivan, L. W., Adams, W. H. & Liu, Y. K. (1977). "Induction of thrombocytopenia by thrombopheresis in man: Patterns of recovery in normal subjects during ethanol ingestion and abstinence." *Blood*, 49, 197-207.

[146] Tan, E. M., Cohen, A. S., Fries, J. F., Masi, A. T., McShane, D. J., Rothfield, N. F., Schaller, J. G., Talal, N. & Winchester, R. J. (1982). "The 1982 revised criteria for the classification of systemic lupus erythematosus." *Arthritis and Rheumatism*, 25, 1271-1277.

[147] Tsai, H. M. & Lian, E. C. Y. (1998). "Antibodies to von Willebrand factor-cleaving protease in acute thrombotic thrombocytopenic purpura." *The New England Journal of Medicine*, 339, 1585-1594.

[148] Vella, M. A., Jenner, C., Betteridge, D. J. & Jowett, N. I. (1988). "Hypothermia-induced thrombocytopenia." *Journal of the Royal Society of Medicine*, 81, 228-229.

[149] Vesely, S. K., George, J. N., Lammle, B., Studt, J. D., Alberio, L., El-Harake, M. A. & Raskob, G. E. (2003). "ADAMTS13 activity in thrombotic thrombocytopenic purpura-hemolytic uremic syndrome: relation to presenting features and clinical outcomes in a prospective cohort of 142 patients." *Blood*, 102, 60-68.

[150] Veyradier, A., Obert, B., Haddad, E., Cloarec, S., Nivet, H., Foulard, M., Lesure, F., Delattre, P., Lakhdari, M., Meyer, D., Girma, J. P. & Loirat, C. (2003). "Severe deficiency of the specific von Willebrand factor-cleaving protease (ADAMTS 13) activity in a subgroup of children with atypical hemolytic uremic syndrome." *The Journal of Pediatrics*, 142, 310-317.

[151] Villalobos, T. J., Adelson, E., Riley, P. A. Jr. & Crosby, W. H. (1958). "A cause of the thrombocytopenia and leukopenia that occur in dogs during deep hypothermia." *Journal of Clinical Investigation*, 37, 1-7.

[152] Vipan, W. H. (1865). "Quinine as a cause of purpura." *Lancet, 2*, 37.

[153] Visentin, G. P., Newman, P. J. & Aster, R. H. (1991). "Characteristics of quinine- and quinidine-induced antibodies specific for platelet glycoproteins IIb and IIIa." *Blood, 77*, 2668-2676.

[154] Wadenvik, H., Denfors, I. & Kutti, J. (1987). "Splenic blood flow and intrasplenic platelet kinetics in relation to spleen volume." *British Journal of Haematology, 67*, 181-185.

[155] Wall, J. E., Buijs-Wilts, M., Arnold, J. T., Wang, W., White, M. M., Jennings, L. K. & Jackson, C. W. (1995). "A flow cytometric assay using mepacrine for study of uptake and release of platelet dense granule contents." *British Journal of Haematology, 89*, 380-385.

[156] Warkentin, T. E. & Smith, J. W. (1997). "The alloimmune thrombocytopenic syndromes." *Transfusion Medicine Reviews., 11*, 296-307.

[157] Watson, S. P., Lowe, G. C., Lordkipanidze, M. & Morgan, N. Y. (2013). "Genotyping and phenotyping of platelet function disorders." *Journal of Thrombosis and Haemostasis*, 2013, 351-363.

[158] White, D. G. (1969). "The dense bodies of human platelets: Inherent electron opacity of serotonin storage organelles." *Blood, 33*, 598-606

[159] Wilde, J. T., Kitchen, S., Kinsey, S., Greaves, M. & Preston, F. E. (1989). "Plasma D-dimer levels and their relationship to serum fibrinogen/fibrin degradation products in hypercoagulable states." *British Journal of Haematology, 71*, 65-70.

[160] Win, N. (1996). "Provision of random-donor platelets (HPA-1a Positive) in neonatal alloimmune thrombocytopenia due to anti HPA-1a alloantibodies." *Vox Sanguinis, 71*, 130-131.

[161] Wolber, E. M. & Jelkmann, W. (2002). "Thrombopoietin: The novel hepatic hormone." *News in Physiological Sciences, 17*, 6-10.

[162] Yujiri, T., Tanaka, Y., Tanaka, M. & Tanizawa, Y. (2009). "Fluctuations in thrombopoietin, immature platelet fraction, and glycocalicin levels in a patient with cyclic thrombocytopenia." *International Journal Hematology, 90*, 429-430.

[163] Zhang, F., Chu, X., Wang, L., Zhu, Y., Li, L., Ma, D., Peng, J. & Hou, M. (2006). "Cell-mediated lysis of autologous platelets in chronic

idiopathic thrombocytopenic purpura." *European Journal of Haematology*, *76*, 427-431.

[164] Zhou, J., Wu, Z., Zhou, Z., Wang, Z., Liu, Y., Huang, X. Y. & Peng, B. (2013). "Efficacy and safety of laparoscopic splenectomy in thrombocytopenia secondary to systemic lupus erythematosus." *Clinical Rheumatology*, *32*, 1131-1138.

BIOGRAPHICAL SKETCHES

M. Manasa

Affiliation: Department of Biotechnology, JAIN (Deemed-to-be University), Bengaluru, Karnataka, India

Education:

- MPhil. in Biotechnology, 2013 from JAIN (Deemed-to-be University), Bengaluru.
- Master's degree in Biotechnology, 2011 from JAIN (Deemed-to-be University), Bengaluru.
- Bachelor's degree [CZBt] 2009 from PES College, Bengaluru

Research and Professional Experience: 6 years of Research experience

Publications from the Last 3 Years:

Manasa, M; Vani, R. L-carnitine as an additive in Tyrode's buffer during platelet storage. *Blood Coagulation and Fibrinolysis.*, 29, 613-621, 2018.

Mithun, M; Rajashekaraiah, V. Platelet storage: Role of Cassia tora Linn. *Asian Journal of Transfusion Science.*, 2018, In Press.

Manasa, K; Vani, R Influence of oxidative stress on stored platelets. *Advances in Hematology.*, 2016. Article ID 4091461, 6 pages, 2016. DOI: 10.1155/2016/4091461.

Manasa, K; Soumya, R; Vani, R. Phytochemicals as potential therapeutics for Thrombocytopenia. *Journal of Thrombosis and Thrombolysis.*, 41, 436-440, 2016.

R. Vani

Affiliation: Department of Biotechnology, JAIN (Deemed-to-be University), Bengaluru, Karnataka, India

Education:

- PhD in Zoology, 2008 from Bangalore University, Bengaluru.
- Master's degree in Zoology, 1997 from Bangalore University.
- Bachelor's degree [CBZ] 1995 from Vijaya College, Bengaluru

Research and Professional Experience: 16 years of Research experience and 12 years of professional experience.

Professional Appointments:

- Associate Professor in Biotechnology, School of Sciences-PG, JAIN (Deemed-to-be University), Bangalore from July 2018 till date. (Courses: Molecular Genetics, Cell Biology, Molecular Biology, Genetic Engineering and Animal Biotechnology).
- Assistant Professor in Biotechnology, School of Sciences-PG, JAIN (Deemed-to-be University), Bangalore from 21st May 2008 till June 2018. (Courses: Molecular Genetics, Cell Biology, Molecular Biology, Genetic Engineering and Animal Biotechnology).
- Lecturer in Biology, VET College, Bangalore for two years from July 2000 to June 2002. (Subjects: Zoology and Botany).

Honors:

- Awarded VGST-SMYSR funding for Seed Money to Young Scientists for Research, in 2013, from the government of Karnataka.
- Qualified the CSIR [NET] JRF test in 2002, ranked one among the top 100 and eligible for Shyam Prasad Mukherjee Fellowship Exam.

Publications from the Last 3 Years:

Manasa M, Vani Rajashekaraiah. "L-carnitine as an additive in Tyrode's buffer during Platelet storage. *Blood Coagul. Fibrin.*, 2018, 29, 613-621.

Manasa M, Vani Rajashekaraiah. Platelet storage: Role of *Cassia tora* Linn. *Asian J. Transfus. Sci.*, 2018. In Press.

Carl, Hsieh; Soumya, Ravikumar; Vani, Rajashekaraiah. Ferric reducing ability of plasma: A potential marker in stored plasma. *Asian J. Transfus. Sci.*, 2018. In Press.

Soumya, R; Vani, R. Vitamin C as a Modulator of Oxidative Stress in Erythrocytes of Stored Blood. *Acta Haematologica Polonica.*, 48, 350-356. 2017.

Carl, H; Vani, R. Influence of L-carnitine on Stored Rat Blood: A study on Plasma. *Turk. J. Hematol.*, 34, 328-333. 2017.

Carl, H; Soumya, R; Srinivas, P; Vani, R. Oxidative Stress in Erythrocytes of Banked ABO Blood. *Hematol.*, 21, 630-634. 2016.

Manasa, K; Soumya, R; Vani, R. Phytochemicals as potential therapeutics for Thrombocytopenia. *J. Thromb. Thrombolysis.*, 41, 436-440. 2016.

Soumya, R; Carl, H; Vani, R. Prospects of curcumin as an additive in storage solutions: a study on erythrocytes. *Turk. J med. Sci.*, 2016, 46, 825-833.

Manasa, K; Vani, R. "Influence of Oxidative Stress on Stored Platelets," *Adv. Hematol*, vol. 2016, Article ID 4091461, 6 pages, 2016. doi:10.1155/2016/4091461.

Soumya, R; Vani, R. Comparison of the Protective Nature of Antioxidants on Stored Erythrocytes. *Appl Med Res*. Online First, 11 Mar, 2016. doi:10.5455/amr.20160309115846.

INDEX

A

acid, 46, 67, 68, 79, 89
acute lymphoblastic leukemia, 94
acute myeloid leukemia, 99
adenosine triphosphate, 4
adhesion, viii, x, 1, 3, 4, 5, 12, 13, 16, 31, 32, 34, 37, 57, 59, 90
adults, 75, 78, 85, 92, 105
aggregation, viii, ix, x, 1, 2, 3, 5, 6, 8, 12, 13, 14, 26, 31, 32, 57, 59, 61, 62, 63, 76, 90, 104
aggregation process, 6
agonist, 13, 32, 74
albumin, viii, 2, 7, 96
alcohol consumption, 70, 99
alcoholism, 70, 92
alkaloids, xi, 58, 75
anaphylactic reactions, 73
anemia, 47, 64, 70
angiogenesis, ix, 10, 12, 14, 30, 32, 33, 34, 44, 46, 48, 52, 53, 54
antibody, viii, 2, 8, 72, 75, 82, 83, 84, 98, 99, 102
anticoagulant, x, 49, 58, 61, 85
antigen, viii, 2, 8, 20, 22, 75, 81, 83, 100, 101
antigen-antibody complexes, viii, 2, 8
antimicrobial peptides (HPAPs), 9
antioxidants, viii, xi, 39, 58, 89, 90, 110
antiphospholipid antibodies, 95, 96
antiphospholipid syndrome, 96, 102
antitumor, 26, 35
antitumor immunity, 26
antiviral therapy, 78
aplastic anemia, 69, 71
apoptosis, 28, 100
arrest, x, 16, 57, 59, 90
ascorbic acid, 75, 89, 98
asymptomatic, 59, 89
atherosclerosis, 12, 16, 62
autoantibodies, 62, 71, 98, 100, 102, 103
autoimmune disease, 12, 73, 77
autologous platelet-rich plasma, v, 44, 54, 55
autosomal recessive, 65, 85

B

bacterial infection, 11, 73

Bernard-Soulier's syndrome, 5
biochemistry, 13
biogenic amines, 7
biomarkers, 26
biopsy, 67
biosynthesis, 45
bladder cancer, 26
bleeding, vii, viii, x, 1, 5, 6, 11, 57, 59, 62, 64, 65, 66, 68, 73, 76, 77, 78, 82, 83, 84, 86, 87, 88, 89, 90, 95, 103
bleeding time, 66, 103
blood, vii, viii, ix, x, 1, 2, 3, 4, 5, 6, 10, 15, 17, 18, 19, 20, 22, 25, 31, 33, 34, 35, 36, 48, 49, 57, 58, 59, 61, 62, 63, 66, 67, 73, 86, 88, 90, 98, 99, 101, 107
blood circulation, 36
blood clot, viii, 1, 2, 4, 6
blood flow, 5, 33, 34, 107
blood smear, x, 58
blood stream, vii, ix, 25, 35
blood transfusion, 19
blood vessels, x, 31, 34, 57, 59, 90
bloodstream, x, 31, 34, 57, 59
bone, x, 10, 31, 53, 57, 61, 62, 69, 70, 95, 97, 100, 102
bone marrow, x, 31, 57, 61, 62, 69, 70, 95, 97, 100, 102
bone marrow biopsy, 62
bone marrow transplant, 95

C

cancer, vii, viii, 25, 26, 29, 30, 31, 32, 33, 34, 35, 36, 37, 38, 39, 41, 45, 91
cancer cells, 30, 32, 34, 35, 37
cancer progression, ix, 26, 32, 33, 34, 37
cardiac surgery, 10
cardiopulmonary bypass, xi, 58
cell cycle, 28, 64
cell line, 32, 33, 64
cell metabolism, 7

cell surface, 64
central nervous system, 59
cerebral hemorrhage, 89
chemokines, ix, 4, 11, 12, 17, 44
chemotaxis, 10
chemotherapy, 69
childhood, 27, 28, 85
children, 72, 85, 94, 101, 106
cholesterol, 3, 74
chronic recurrent, 95
circulation, x, 5, 12, 33, 34, 57, 69, 82
classification, 17, 71, 97, 102, 106
coagulation process, vii, viii, 1, 4
coagulopathy, 94, 96, 97, 104
collagen, ix, 3, 4, 5, 7, 10, 14, 16, 31, 34, 44, 45, 46, 49, 52, 53, 66, 67, 94
complete blood count, 63
complex GPIb/IX/V, 5
complications, viii, 1, 45, 76, 78, 90, 92
connective tissue, 3, 4, 5, 10, 45, 52
consumption, 61, 68, 70, 71, 76, 78, 86, 105
contraceptives, 68, 89
controlled trials, 50
coronary artery disease, 12
cryopreservation, 21
cutaneous melanoma, v, vii, viii, 25, 26, 27, 28, 35, 37, 39
cyclophosphamide, xi, 58, 77, 92
cyclosporine, xi, 58, 85, 89
cytokines, 11, 34, 46, 47, 64
cytomegalovirus, 70
cytometry, 67, 81, 83, 97

D

debridement, ix, 44, 47, 48
defects, vii, viii, 1, 10, 13, 66, 67
deficiencies, vii, viii, 1, 5, 6, 67, 85
deficiency, 46, 70, 75, 84, 85, 91, 97, 100, 106
dengue, 89, 90, 93, 99, 106

dengue fever, 93, 99, 106
dengue hemorrhagic fever, 89
destruction, x, 58, 61, 71, 72, 73, 75, 76, 77, 78, 81, 84, 86, 87, 88
detection, 83, 96, 97
diabetic patients, 45
diagnostic markers, 15
DIC, 77, 86, 87, 88, 96, 99, 104
difficult-to-heal venous leg ulcers, 44
disease progression, 34
diseases, 8, 45, 69
disorder, x, 27, 45, 57, 65, 69, 85, 87, 88, 90
disseminated intravascular coagulation, 77, 96
distribution, x, xi, 19, 58, 61
drugs, xi, 58, 68, 77, 78, 79, 80, 81, 89

E

Ehlers-Danlos syndrome, 94
endothelial cells, viii, 2, 4, 5, 7, 10, 11, 12, 13, 14, 33, 34, 46
endothelium, x, 3, 7, 12, 13, 57, 59, 90
environment, 3, 8
environmental factors, 80
enzyme-linked immunosorbent assay, 68
epidermis, 26, 27, 29, 44
erythrocytes, 49, 87, 110
E-selectin, 12
estrogen, 74
ethanol, 91, 106
ethnic groups, 27
ethylene, 98, 103
etiology, 27, 61, 88
exposure, 5, 27, 28, 29, 31, 81, 82
extracellular matrix, 31, 33
extravasation, vii, viii, 1, 2, 30, 32, 34

F

family history, 62, 65

family members, 65
fibrin, vii, viii, 1, 2, 8, 9, 14, 17, 32, 33, 52, 68, 84, 86, 107
fibrin degradation products, 86, 107
fibrinogen, 3, 6, 8, 9, 32, 84, 86, 107
fibrinolytic, 84, 85, 96
fibroblast growth factor, 52
fibroblasts, 4, 10, 46, 48
formation, vii, viii, ix, x, 1, 2, 3, 6, 8, 12, 13, 25, 31, 33, 34, 35, 36, 45, 46, 50, 52, 53, 57, 59, 86, 90, 91
fresh frozen plasma, 88

G

gastrointestinal bleeding, 77
gel, viii, 2, 9, 14, 17, 47, 49, 50, 54
gene therapy, 69
genes, 5, 28, 64, 68
genetic defect, x, 58
genitourinary tract, 59
gingival, 64, 68, 74, 89
glomerulonephritis, 8
glucocorticoids, xi, 58, 77, 89
glycoproteins, 3, 31, 76, 107
graduate students, 18
granules, 3, 4, 6, 7, 9, 11, 13, 31, 32, 33, 67
growth, viii, ix, 2, 9, 10, 26, 29, 30, 31, 33, 44, 45, 46, 47, 48, 52, 53, 74, 94, 104
growth factors, viii, ix, 2, 9, 10, 34, 44, 45, 46, 47, 48, 52, 54, 94

H

hard tissues, viii, 2, 9, 10
headache, 73, 74, 75
healing, vii, viii, ix, 2, 9, 10, 44, 46, 47, 49, 50, 51, 52, 54
hemodialysis, xi, 58
hemolytic anemia, 83, 84

hemolytic uremic syndrome, 77, 84, 92, 96, 106
hemorrhage, 65, 83, 86
hemostasis, vii, viii, 1, 2, 3, 4, 6, 9, 13, 30, 59, 92, 93, 98, 99, 101
hepatic failure, 86
hepatitis, 70, 73, 75, 78, 103, 105
human, x, 3, 9, 14, 15, 22, 26, 34, 52, 57, 59, 70, 84, 94, 107
human immunodeficiency virus, 70, 94
hypersplenism, xi, 58, 77, 98, 101
hypertension, 73, 74
hypodermis, 45
hypoplasia, 77, 89
hypothermia, xi, 58, 87, 95, 103, 106

I

idiopathic, 65, 88, 93, 99, 100, 104, 105, 108
idiopathic thrombocytopenic purpura, 65, 93, 99, 100, 105, 108
immune defense, 47
immune function, 26
immune modulation, 73
immune response, 11, 74
immune system, 11, 31, 71
immunoglobulin, 77, 101
immunoglobulins, viii, 2, 7
immunosuppression, 74
immunosuppressive drugs, 77
immunotherapy, 26, 35, 36
in vitro, 9, 32, 52, 61
in vivo, 14, 15, 16, 52, 53
individuals, vii, viii, 25, 29, 31, 67, 102, 103
infection, ix, 11, 44, 45, 47, 71, 75, 78, 85, 89, 101, 105
infectious mononucleosis, 78
inflammation, vii, ix, 9, 10, 11, 12, 14, 16, 25, 35, 44, 45, 46, 53, 59
inflammatory cells, ix, 44

inflammatory mediators, 11, 12
inflammatory processes, viii, 2, 11, 12, 13, 15, 31
injury, iv, vii, viii, x, 1, 2, 4, 8, 57, 59, 90
integrin, 6, 13, 16, 31, 81, 87
intravenous immunoglobulins, xi, 58

K

keratinocyte, 27, 53
keratinocytes, 46

L

leukocytes, viii, ix, 2, 5, 11, 12, 13, 16, 44, 49
leukopenia, 64, 70, 106
liver, 62, 70, 71, 74, 75, 78, 87, 100, 101
liver cirrhosis, 70, 78
liver disease, 87, 100, 101
liver enzymes, 74
lupus erythematosus, 12, 104
lymph node, 30, 73
lymphatic system, 30
lymphocytes, ix, 11, 44, 75

M

macrophages, ix, 4, 10, 11, 44, 47, 53, 59, 73, 75, 87
malaria, 71, 103
malignancy, 32, 62, 74
malignant cells, 30, 69
malignant melanoma, 30
marrow, 59, 61, 67, 69, 70, 71, 73, 74, 75, 77, 78, 102
medical, ix, 43, 44
medication, ix, 44, 89
medicine, 17, 18, 99
megakaryocyte, 71, 74, 92, 100, 105

melanoma, vii, viii, 25, 26, 27, 28, 29, 30, 35, 36, 37
membranes, 3, 4, 34, 76
menorrhagia, 65, 68, 89
metabolic changes, 30
metabolism, viii, 2, 17, 30
metabolites, 81
metastasis, vii, ix, 26, 30, 32, 33, 34, 35, 36
methemoglobinemia, 75
methodology, 17
methylprednisolone, 73, 82, 85
microcrystalline, 49
microorganisms, 9
microparticles, 7, 12, 15, 33, 35, 85
microscopy, 67
migraine headache, 7
migration, ix, 3, 16, 26, 34, 44, 46, 47, 53
molecular pathology, 68
molecular weight, 3, 80
molecules, vii, ix, 6, 12, 15, 16, 25, 32, 35, 44, 45, 64, 82
monoclonal antibody, 81, 82, 100, 105
morphology, x, 13, 58, 61, 63, 64, 90, 97
mortality, 26, 35, 83, 85
mortality rate, 85
myelodysplasia, 70, 91
myelodysplastic syndromes, 97
myocardial infarction, 12

N

neutrophils, 11, 12, 16
non-Hodgkin's lymphoma, 69
non-neoplastic diseases, 69
nutritional deficiencies, x, 58

O

organ, 73, 74, 77, 86, 87
organelles, 27, 107

P

pancreatic cancer, 26
parasitic infection, 78
partial thromboplastin time, 84, 86
pathogenesis, x, 12, 15, 58, 61, 71, 75, 76, 85, 91, 96, 104, 105
pathophysiological, viii, ix, 2, 44, 45
pathophysiology, 15, 71, 90, 98
peripheral blood, 20, 47, 61, 62
peripheral neuropathy, 74, 75
phagocytic cells, 74
phagocytosis, 15, 73
phosphatidylethanolamine, 7
phosphatidylserine, 7
phospholipids, viii, 1, 3, 7
platelet activating factor, 5
platelet activation, ix, 3, 5, 7, 9, 16, 26, 32, 33, 34, 37, 87
platelet aggregation, viii, 2, 3, 6, 8, 13, 14, 32, 61, 62, 63, 67, 76
platelet concentrates (PCs), 9, 17, 20, 22, 83
platelet count, x, 16, 26, 35, 36, 57, 59, 64, 66, 68, 70, 73, 74, 75, 76, 78, 82, 83, 87, 88, 89, 90, 105, 106
platelet derived growth factor (PDGF), ix, 4, 10, 44, 46, 47, 52
platelet disorders, x, 58, 61, 64, 65, 69, 92, 93, 104
platelet function disorders, 90, 97, 107
platelet gel, viii, 2, 9, 10, 14, 17, 54
platelet granules, 3, 4, 6, 11, 67
platelet plug, vii, viii, 1, 2, 8
platelet release reaction, 3, 6, 13, 40
platelet-rich plasma (PRP), v, ix, 9, 14, 43, 44, 47, 48, 49, 50, 51, 53, 54
platelets, v, vii, viii, ix, x, 1, 2, 3, 4, 5, 6, 7, 8, 9, 11, 12, 13, 14, 15, 16, 17, 25, 26, 30, 31, 32, 33, 34, 35, 36, 37, 38, 39, 40, 41, 47, 52, 57, 58, 59, 62, 64, 65, 67, 68, 70, 71, 73, 75, 76, 78, 80, 81, 82, 83, 84,

86, 87, 88, 90, 91, 93, 94, 95, 97, 98, 102, 104, 107, 109, 110
polycythemia vera, 105
population, ix, 36, 43, 44
portal hypertension, 87
pregnancy, x, 58, 65, 74, 76, 100
preparation, iv, 17, 19, 20, 48, 49
primary tumor, vii, ix, 25, 31, 34, 35
pro-inflammatory cytokines, 11
proliferation, ix, 9, 10, 11, 26, 28, 33, 44, 45, 46, 47, 48, 52, 64, 74
prostaglandin E1, 8
proteins, 3, 4, 5, 32, 33, 64, 68, 82, 85
P-selectin, 7, 12, 16, 38
pseudopodia, 6, 8
purpura, 64, 72, 101, 102, 104, 107

R

receptor, xi, 3, 5, 6, 7, 15, 16, 27, 58, 74, 75, 78, 80, 81, 88, 93
receptors, 3, 5, 6, 11, 15, 17, 32, 64, 73, 74
reconstruction, 9, 10, 50
recovery, 81, 90, 106
regeneration, vii, x, 10, 44, 45, 46, 47, 48, 50
regenerative medicine, 9
renal dysfunction, 74, 85
response, viii, ix, 2, 9, 15, 44, 50, 75, 83, 85, 86
risk, 12, 27, 28, 45, 69, 71, 73, 74, 76, 78, 82, 83, 84, 103
risk factors, 27, 28, 84

S

secrete, 11, 32, 34, 48
secretion, x, 12, 13, 46, 47, 57, 59, 90
serotonin, 4, 5, 6, 7, 107
serum, 46, 52, 97, 107
services, iv

side effects, 87, 90
signaling pathway, 28
skin, vii, x, 14, 26, 27, 29, 44, 45, 46, 47, 48, 50, 59, 75
skin cancer, 26, 29
smooth muscle, 10, 52
smooth muscle cells, 10, 52
spleen, 59, 61, 73, 75, 78, 86, 87, 91, 107
splenomegaly, xi, 58, 70, 87
stem cells, 18, 64, 70, 71
stimulation, 10, 46, 47, 53
storage, 17, 18, 20, 67, 97, 107, 108, 110
storage pool disease, 67
stress, x, 5, 33, 34, 57, 59
suppression, 26, 70, 71, 74
surgical technique, 9
survival, 14, 26, 30, 35, 36, 37, 88, 93
susceptibility, 27, 28, 80
symptoms, 8, 78, 85, 87, 89
syndrome, 5, 21, 64, 66, 67, 70, 76, 90, 95, 96, 97, 102, 104
synthesis, ix, 4, 7, 9, 10, 44, 46, 52, 53, 70, 73
systemic lupus erythematosus, 16, 76, 91, 92, 93, 94, 97, 101, 103, 106, 108

T

temperature, 63, 87, 88
therapeutics, v, 57, 58, 73, 89, 90, 109, 110
therapy, ix, 8, 44, 45, 47, 50, 70, 73, 75, 78, 85, 89, 94, 95, 99, 101, 104
thrombin, 5, 6, 7, 9, 32, 33, 35
thrombocytopenia, v, vii, x, xi, 11, 15, 47, 57, 58, 59, 60, 61, 63, 64, 67, 69, 70, 71, 72, 76, 77, 78, 79, 80, 81, 82, 83, 84, 85, 86, 87, 88, 89, 90, 91, 92, 93, 94, 95, 96, 97, 98, 99, 100, 101, 102, 103, 104, 105, 106, 107, 108, 109, 110
thrombocytopenic purpura, 72, 77, 78, 92, 93, 95, 96, 98, 101, 103, 104, 106

thrombocytosis, 35, 74, 88
thrombopoietin, xi, 58, 68, 97, 107
thrombosis, 73, 76, 78, 81
thrombus, 33
tissue, vii, x, 4, 9, 11, 18, 19, 30, 31, 32, 34, 44, 45, 46, 47, 48, 50, 52, 75, 77, 86
toll like receptors (TLRs), 11
transcription, 64, 74
transcription factors, 64
transfusion, 17, 18, 62, 83, 84, 88, 93, 97, 101, 102, 105
trauma, 9, 11, 14, 45, 59, 65, 86, 88, 93, 94, 96
treatment, viii, ix, x, xi, 8, 11, 20, 21, 26, 30, 35, 44, 46, 47, 50, 52, 53, 54, 58, 62, 68, 73, 76, 77, 78, 83, 85, 86, 88, 90, 92, 98, 101, 104, 105
treatment methods, 46, 76
tumor, ix, 11, 15, 26, 27, 29, 30, 32, 33, 34, 59
tumor cells, ix, 26, 32, 33, 34
tumor growth, 33, 59
tumor metastasis, 30
tumor necrosis factor, 11, 15

U

ulcer, vii, x, 44, 47, 48, 49, 50, 53, 54
ultrastructure, 67

V

vascular endothelial cells, 4, 11
vascular endothelial growth factor (VEGF), 46
vascular system, ix, 44, 45
vascular wall, 7
vasculature, 34, 59
vasoactive substances, 8
vessels, 5, 8, 10, 30, 34, 59
viral infection, xi, 58, 69, 70, 72
vitamin B1, 70
vitamin B12, 70
vitamin B12 deficiency, 70
vitamin C, 89
vitamin E, 89, 93
vW factor, 5

W

withdrawal, xi, 58
worldwide, vii, viii, ix, 25, 44
wound healing, viii, 2, 9, 10, 49, 50, 51, 52, 53, 59

Related Nova Publications

CALMODULIN: STRUCTURE, MECHANISMS AND FUNCTIONS

EDITOR: Vahid Ohme

SERIES: Cell Biology Research Progress

BOOK DESCRIPTION: In *Calmodulin: Structure, Mechanisms and Functions*, the authors consider small and poorly-studied groups of plant calcium-dependent protein kinases that directly interact with calmodulin molecules.

SOFTCOVER ISBN: 978-1-53614-948-7
RETAIL PRICE: $82

FLAGELLA AND CILIA: TYPES, STRUCTURE AND FUNCTIONS

EDITOR: Rustem E. Uzbekov

SERIES: Cell Biology Research Progress

BOOK DESCRIPTION: Motility is an inherent property of living organisms, both unicellular and multicellular. One of the principal mechanisms of cell motility is the use of peculiar biological engines – flagella and cilia. These types of movers already appear in prokaryotic cells. However, despite the similar function, bacteria flagellum and eukaryote flagella have fundamentally different structures.

SOFTCOVER ISBN: 978-1-53614-333-1
RETAIL PRICE: $95

To see a complete list of Nova publications, please visit our website at www.novapublishers.com

Related Nova Publications

MITOGEN-ACTIVATED PROTEIN KINASES (MAPKs): ACTIVATION, FUNCTIONS AND REGULATION

EDITOR: Charles K. Hester

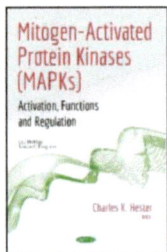

SERIES: Cell Biology Research Progress

BOOK DESCRIPTION: *Mitogen-Activated Protein Kinases (MAPKs): Activation, Functions and Regulation* opens with a summary of the present knowledge about MAPK, with special emphasis on p38 and c-Jun N-terminal kinase. The authors focus on how these signaling pathways are engaged during some infections with intracellular parasites.

SOFTCOVER ISBN: 978-1-53616-138-0
RETAIL PRICE: $69

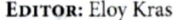

BETA-GALACTOSIDASE: PROPERTIES, STRUCTURE AND FUNCTIONS

EDITOR: Eloy Kras

SERIES: Cell Biology Research Progress

BOOK DESCRIPTION: In *Beta-Galactosidase: Properties, Structure and Functions*, the authors discuss the main microorganisms that produce β-galactosidase, the characteristics of the culture media, bioprocessing parameters, the most relevant downstream steps used in the recovery of microbial β-galactosidase, as well as the main immobilization techniques.

SOFTCOVER ISBN: 978-1-53615-605-8
RETAIL PRICE: $95

To see a complete list of Nova publications, please visit our website at www.novapublishers.com